P9-EDH-600

WHAT YOUR FOOD ATE

ALSO BY DAVID R. MONTGOMERY

Growing a Revolution

The Rocks Don't Lie

Dirt

King of Fish

ALSO BY DAVID R. MONTGOMERY AND ANNE BIKLÉ

The Hidden Half of Nature

WHAT YOUR FOOD ATE

How to Heal Our Land and Reclaim Our Health

DAVID R. MONTGOMERY
AND ANNE BIKLÉ

W. W. NORTON & COMPANY
Independent Publishers Since 1923

Printed in the United States of America
First Edition

Manufacturing by Lakeside Book Company
Book design by Daniel Lagin
Production manager: Lauren Abbate

ISBN: 978-1-324-00453-0

W. W. Norton & Company, Inc.
500 Fifth Avenue, New York, N.Y. 10110
www.wwnorton.com

W. W. Norton & Company Ltd.
15 Carlisle Street, London W1D 3BS

1 2 3 4 5 6 7 8 9 0

For the farmers, scientists,
and eaters around the world
dedicated to bringing life back to the soil.

CONTENTS

ANIMAL

PEOPLE

INTRODUCTION

Most of us don't think about the soil when we sit down for a meal with friends or family. But we should. We all know that a fresh pear is healthier than a pile of salty potato chips, but it is all too easy to lose sight of another dimension to our health: what's in that pear and how it got there—how we grow our food.

In a world awash in clickbait headlines about what to eat and what to avoid, few know that the typical grocery store carrot now contains less zinc than those our great-grandmothers served their kids, or that the beef sizzling on the barbeque probably packs a lot less iron than what our grandparents ate as children. Reports of troubling nutrient declines span the human diet, from fruits and vegetables to grains, meats, and dairy. And while what we mostly hear in arguments over the future of agriculture centers on differences between organic and conventional farming, the real story is not that simple—and far more interesting.

Over the past century farming practices have changed food in ways that reduced levels of beneficial compounds in our diet, ranging from those in fruits and vegetables that help prevent cancer to fats in meat and dairy that modulate inflammation. Too many of us have never heard of fats that are good for us or know that our health suffers when we don't get enough plant-made phytochemicals in our diet. Yet when it comes to lifelong health, an adequate supply of such compounds appears as fundamental as getting enough exercise.

There is no shortage of opinions about what we should eat. People argue endlessly, for example, over whether we should eat less meat, more meat, no meat, or meat that isn't meat. What's typically missing from the framing of dietary choices is *how* we grow what we eat. The way we raise our crops and livestock proves as important as what we choose to eat.

We're still learning how soil life helps plants grow and influences the things that end up in their bodies and thus in our bodies. What role does better soil play in growing a better peach or carrot? If we want the healthiest food, is organic farming good enough? To answer such questions we need to examine the long-running relationships that all plants develop with a vast menagerie of soil life. You may have heard about the importance of our microbiome, particularly the gut microbiota that dwell in our colon and transform what lands there into compounds that benefit our health. Similarly, the tiniest soil dwellers influence crop health. And what's true in the human body holds true for soil too. Routinely killing off or disrupting microbial partnerships rarely benefits the host.

Soil is the starting point for foods that come from the land, and a groundswell of evidence, observations, and research points to an underappreciated factor contributing to food quality—the health of the soil on farms and ranches. As we'll share in the pages ahead, a substantial body of evidence shows that the state of the land influences the health of crops, for better and worse. And the nutritional quality of the pasture, crops, or rangeland that livestock eat strongly influences their health. But does soil health affect human health as well? How much does what your food ate matter to your health? Like beads on a string, we can connect the way farmers treat their soil, grow crops, and feed their livestock to what fills our plates, glasses, and bodies.

How we treat the land affects how it, in turn, treats us. Key pillars of today's conventional agriculture—routine plowing and liberal use of chemical fertilizers to grow relatively few high-yielding crops—helped drive the twin problems of degraded soil health and declining nutrients in food. Over the past century our food lost mineral elements we need in trace amounts, like copper and zinc, and others we need a lot more of, like calcium and magnesium. Today, you'd have to eat several apples a day to keep that proverbial doctor away.

These changes didn't happen by design as much as by oversight and a single-minded focus on quantity. We've learned a lot in recent decades about the ways crops acquire or get shorted on minerals and other nutrients and compounds important for our health. Out-of-sight, belowground communities of symbiotic soil bacteria and fungi work as a nutrient acquisition and intelligence service for plants. And it turns out that farming practices shape these communities and thus the types and amounts of health-promoting minerals, fats, and phytochemicals in our crops, meat, and dairy—and ultimately in our bodies.* But the go-to practices of mainstream agriculture disrupt the soil microbiome, unleashing modern farming's chemical treadmill. The more farmers come to rely on synthetic inputs instead of beneficial soil life, the more they need the former—and lose the latter.

Fortunately, as we'll see, there are practical, cost-effective ways for farmers to cultivate soil life and harvest healthier food. But doing so requires thinking differently about the soil. In coming chapters you'll meet visionaries who long ago foresaw that losing soil life would seed a downward spiral toward less nutritious food and less healthy people, and you'll read about innovative farmers bringing their land back to life and nutrition back to their harvests. This is a story about the way health flows to us from the land. It matters how we farm, and we needn't

* Phytochemicals are compounds plants make to help them defend themselves, enhance their growth, and communicate.

choose between what's healthy for the environment and for people. We can feed ourselves in ways that build and safeguard the health of both the land and our bodies.

Regenerative farming practices can help stop soil erosion, enhance soil health, and build up soil organic matter (soil carbon). They are catching on among American farmers, allowing them to slash spending on fuel, fertilizers, and pesticides. In effect, farmers who adopt regenerative practices find they can rely on biology to replace most of what they used to buy from agrochemical suppliers. At the same time, consumer demand for more sustainable products incentivizes a growing list of companies integrating regenerative farms into their supply chains. But do these products really prove better for our health too? Does science back up claims of better food from healthy, fertile soil, and if so, what could regenerative agriculture mean for human nutrition—and public health?

We tend to think of diet-related ailments as arising from deficiencies in particular nutrients. Take scurvy, for example. If you start consuming vitamin C–rich citrus, scurvy clears right up, as British naval surgeon James Lind demonstrated in a famous 1749 medical trial aboard HMS *Salisbury*. Some cures are this simple. But curing other types of ailments and maintaining good health in the first place are far more complex. And on this point, the mix of foods in the human diet matters because phytochemicals, minerals, fats, and other compounds in food interact synergistically. And the fact that farming practices influence all of them leaves us with an unsettling question. How good, really, is modern agriculture for our health?

Of course, we're not arguing that food grown in degraded soil is the cause of all the worrisome health trends humanity faces. To be sure, our health—good, bad, and in between—reflects a constellation of interacting factors, from our genes, diet, physical activity level, and microbiome to personal and maternal exposure to toxins and pathogens in the environments where we live, work, and play. Also, the dramatic rise in the onset and types of chronic diseases over the past century is partly attributable to more comprehensive diagnosis and monitoring. But given recent advances in our understanding of the human immune system, as well as in our healing and aging processes, it's time to take a fresh look at whether mainstream agricultural practices fill our crops and livestock with all our bodies need to support robust health.

A key reason so few of us are aware of nutrient declines in food is that society remains fixated on an ancient challenge we've almost, if perhaps only temporarily, solved—growing enough to feed us all. While access to sufficient food remains a conspicuous problem for far too many, hidden hunger lurks beneath the surface of the modern sea of cheap, plentiful calories flooding the westernized world. Focusing on yields, like bushels per acre, pounds of an animal, and gallons of milk, is laudable. It helps fill bellies and pay farmers. But quantities harvested tell us nothing about whether those foods contain enough of the right mix of nutrients to help us grow when young and imbue us with health throughout life, especially as we age.

Honed through the slow-burning fires of evolution, our bodies developed an innate nutritional wisdom. Deeply embedded in our biology, it worked like a compass guiding us toward a healthy diet. Along the way, our senses of touch, smell, and taste helped answer the question of what to eat or not to eat. About half our brain is dedicated to processing, filtering, and interpreting what we see. And what food looked and tasted like contained clues we relied on to guide us to nutrition.

These days, however, relying on our nutritional wisdom to judge the quality of food in a modern grocery store is a fraught proposition. We've all experienced the disappointment of biting into a gorgeous, yet mealy peach or a shiny, mushy apple. Modern marketing and the ubiquity of highly processed foods further confuse and misdirect our inherent nutritional wisdom. Like candy for the eye, highly processed foods in colorful packages adorned with mouthwatering images lure us to overload on seductively flavored yet empty calories. In addition, crop breeders generally favor selecting for yield, harvest convenience, shipping, and storage rather than flavor and nutrition. No wonder many of us increasingly question what's in our food, what might be missing, and why.

So what's a health-conscious consumer to do? Does an organic diet provide an easy answer? Not necessarily. As we'll see, one can't always rely on the familiar green-and-white organic label to deliver nutrient-rich food. Most organic farmers are not so different from conventional farmers when it comes to agriculture's most iconic act—plowing. And running a plow through a living ecosystem is much like having a tornado strike your home. In the case of soil-dwelling organisms that plants rely on, they suffer and their house falls apart. In addition, other practices that have been fought over but are now allowed under the organic label—like soil-less hydroponic production and confined dairy and chicken operations—do not necessarily deliver the nutrition or animal welfare that most consumers equate with organic food.

There are also many reasons beyond the kitchen to care about how we grow our food. After all, the long arm of modern agriculture stretches beyond farms, contributing to downstream water pollution and climate change. Moreover, the health and well-being of those who work on farms remain social justice issues that should concern us all. Adopting farming practices that are better for the land at large—and soil health specifically—would help address these related problems as well.

Must we choose between quantity and quality, or can we grow enough to nourish us all? Could healing sick soil through regenerative agriculture restore fertility to the land, help heal an ailing populace, and enhance our quality of life? Defined broadly, regenerative farming practices are those that build soil organic matter and support soil life in ways that enhance and maintain fertility over time.

Our journey asking questions and writing about connections between people, agriculture, and health began three books ago. The first, *Dirt: The Erosion of Civilizations*, investigates the role of soil erosion and degradation in limiting the life span of past societies. *The Hidden Half of Nature* shares the new science of microbiomes in plants and people, revealing the striking parallels between the soil and the human gut that point to much-needed changes in agriculture and medicine. The third, *Growing a Revolution*, takes this new science and the lessons of history to explore how regenerative farming can rebuild soil health, stash carbon belowground, and greatly reduce agrochemical use. While these prior books respectively map onto history, science, and solutions, you can read them in any order—or start with this one.

Through the eyes and experience of a geologist (Dave) and a biologist (Anne), we'll delve into how farming practices ripple from soil to plant to animal to us. Navigating the tangled history and science behind our relationship with the land, we reexamine beliefs about health, medicine, our bodies, and their intersection with food and farming. Along the way, we'll explore fields from agronomy and microbiology to environmental and nutritional science, as well as bovine and human biology. Insights linking these fields show how and why regenerative farming practices can heal the soil, bring life back to the land, and grow the nutrient-dense and flavor-packed foods we need to reclaim our health. Once you understand the connections, it becomes clear that for plants, animals, and people alike, the roots of health grow from healthy soil.

WHAT YOUR FOOD ATE

THE EDIBLE PUZZLE

It is health that is real wealth.

—MAHATMA GANDHI

Both hope and hype swirl around using food as medicine today. Yet the more we learn, the more a healthy diet comes into focus as a solid foundation for preventing—and potentially treating—chronic diseases. But this isn't really a new idea. Around 400 BCE the Greek physician generally credited with bringing reason into medicine advocated a nutritious diet as the best defense against sickness and disease. In his list of questions that a doctor should always ask, Hippocrates included where a patient's food came from. He saw what one ate and the nature of the soil it grew in as essential pieces in the puzzle of health—good or bad.

As Western medicine sidelined Hippocrates's view, it proved remarkably adept at developing effective treatments and cures for acute maladies and many of the worst infectious diseases. Our grandparents feared scourges like typhoid and scarlet fever that we know now only as haunting names.

But the nature of illness radically shifted over time as public health measures reduced the incidence of infectious diseases through water-supply treatment, antibiotics, and vaccines. Today, however, illnesses unfamiliar to past generations plague us more than ever. We improved our medical prowess only to face escalating rates of chronic diseases and autoimmune conditions that now afflict many Americans.

What happened? Genetics, of course, helps explain why some people are more, or less, prone to particular maladies. But it doesn't fully account for the rapid rise of modern chronic diseases. Our genes don't change that fast.

What did? What we eat and how we grow it. Over the course of the twentieth century we shifted from a diet of whole foods to one low in fiber, rich in sugar and heavily processed food, and grown with a lot of synthetic fertilizer and agrochemicals. It increasingly appears that, although unintended, these dietary and farming changes undermined public health. As agriculture chased calories over nutritional quality, food processing radically reshaped food, and medical research focused on developing new treatments and drugs over investigating the role of diet in preventing disease.

Shifting Wisdom

Scan the health and diet section of any bookstore or peruse Amazon's top-selling nutrition titles, and you may be left wondering what's actually OK to eat. Books about diets and dieting generally push the idea that the path to good health follows eating the "right" foods and avoiding the "wrong" foods. Diets like paleo, keto, and vegan incompatibly emphasize eating certain foods and going heavy or light on carbs, protein, or fats. Across the board, dietary advice typically focuses on what and how much to eat, with remarkably little attention paid to how farming practices influence the nutritional quality of food and whether the "right foods" pack the nutrients they once had.

Just as what we eat has changed over time, so too has our understanding of what makes for a healthy diet. Nutritional wisdom shifted as we learned there is more to diet than just calories and the simple distinction of fats, carbs, and protein. It turns out that fats are not really all that fattening, and too much sugar messes with our metabolism and gets stored as—you guessed it—fat. Because genes and microbiomes differ between individuals and nutritional needs change over a lifetime, there is no one best diet for all people. Despite this inherent variability, diet remains the foundation for good health.

New science about the human microbiome is also part of the evolving understanding around the influence of diet on human health and well-being. Our gut microbes dine on fiber and other compounds in food, producing an array of metabolites, many of which our bodies need. That roughly 40 percent of the molecules circulating through our bodies arise directly or indirectly from our gut microbiota adds a new dimension linking diet to health. As we'll see, what our gut microbiota eat helps explain the now-standard nutritional wisdom of a diet loaded with fruits and vegetables.

Millions of years ago, our earliest primate ancestors ate a fruit heavy diet. Later pre-humans chewed all day on fibrous roots and plants, prepping them for microbiota in the lower reaches of the gut to further digest. Over the ages, we evolved and thrived together with our tiny tenants, giving them food and shelter in exchange for nourishment and microbial metabolites that enhanced our health.

When our pre-human ancestors started using fire to cook food, their bodies began changing. Primatologist Richard Wrangham argues that the adoption of cooking around 1.8 million years ago led to the development of *Homo erectus*, a theory he bases on changes observed in pre-human anatomy during that time. Though the earliest date remains controversial, most researchers seem to agree that our ancestors controlled fire by around half a million years ago. Over countless generations the more efficient nutrient uptake that came with the shift from raw to cooked foods led to reallocating energy previously used for digestion to grow and power our brain. Cooking also helped our early ancestors integrate meat into their diet—when they could get it.

Another major change in the human diet occurred around 10,000 years ago. In the postglacial world we transitioned from hunting and gathering to growing food and raising animals—to agriculture. Herding animals ensured a steady supply of meat that didn't involve a daylong, perilous journey tracking, killing, and hauling. And livestock came with the bonus of milk to consume fresh on the spot or ferment into yogurt or cheese, providing short-term storage of nutritious, flavorful food. Livestock also produced manure—a powerful natural fertilizer that helped crops grow.

Herding and farming fed more people more reliably than hunting and gathering. The crops that dramatically reshaped our postglacial diet were not leafy greens but calorie- and nutrient-rich grains that stored well and provided year-round meals. This meant few could feed many, opening the door for occupational specialization and the rise of large, increasingly complex civilizations.

The most recent, modern shift in what we eat and how we grow it changed the game, especially for our gut microbiota. In reducing our ration of fermentable fiber, we reduced theirs, not fully understanding that this scuttled the production of microbial metabolites our bodies evolved to rely on. In addition, introducing novel additives and chemicals into food impacts the human microbiome (and us) in ways we are only now realizing. Likewise, modern farming methods changed relationships between soil life, crops, and livestock. How much has all this affected our health? More than we bargained for.

The Yield Trap

Talk to farmers for long enough, and sooner or later you'll hear something like, "If it's not in the bin, it doesn't matter." What they mean is that bushels and pounds sold pay the bills. And while tonnage that translates into cheap, abundant food sounds like good news for a hungry world, it is not necessarily good for our health.

So how did we reach this awkward point—astounding harvests of lower nutritional quality? After the Second World War, farmers increasingly relied on mechanized plowing, chemical fertilizers, and pesticides to reap high crop yields. The effects were seductive and addictive—greater yields despite degraded fields. Society equated this edible bounty with human health for good reason. Time and again throughout history famine reduced our numbers. The high yields of modern agriculture seemed miraculous against this backdrop given our rapidly growing population. Today the Western world stands lifetimes removed from mass starvation. Globally, we grow enough to feed everyone if distribution were equitable.

Yet it is well known in agricultural circles that breeding crops for high yields and easier harvesting, processing, and shipping reduced nutritional quality. It is no secret that processing wheat into white flour strips the grain of nutrition. And it doesn't help if nutrients do not get into food in the first place.

Our tendency to believe more is better—to want a larger apple, a bigger tomato, super-sized wheat grains, and more milk per cow—led us to see calories as a stand-in for nutrition. But we neglected to pay attention to what may be missing from those calories even as we learned there is more to health than sufficient calories. Many of the chronic diseases and ailments that plague us today are linked to diet.

Our knowledge of what makes for a good diet expanded beyond a full belly once we discovered that noncaloric things in food like vitamins and minerals also play key roles in good health. Vitamins and certain minerals are often referred to as "micronutrients," reflecting that we need them in small amounts. And so although we may not need much of the mineral element copper, for example, that little bit is central to the workings of our immune system. Yet we can't eat ore, raw metal, or pennies. We have to get copper in our food. Without it or other mineral micronutrients, such as zinc, iron, manganese, and selenium, health suffers.

Globally, micronutrient malnutrition is now more common than inadequate calories. Iron deficiency afflicts an estimated two billion people. Zinc deficiency affects at least a fifth of humanity and is particularly high in sub-Saharan Africa and South Asia. Almost a third of the U.S. population is deficient in at least one vitamin, and almost a quarter lacks sufficient iron. A high proportion of children and adolescents rely on foods that have minerals added. Yet even with such fortified food and supplements, many Americans do not consume recommended levels of vitamins and minerals. For example, a majority of us don't get enough magnesium. Less than a third of Americans get enough vitamin D and the problem is worse among children and teens. Unfortunately, most Americans live with the hidden hunger of a diet rich in calories but poor in micronutrients and phytochemicals. Far too many of us remain poorly nourished despite eating more than enough food.

By the second half of the twentieth century, heart disease, hypertension, and type 2 diabetes became common life-shortening health conditions. Seven out of ten Americans now die of chronic diseases, and around half of us live with at least one such disease. Once-rare autoimmune disorders fill out a slate of modern ailments that plague the Western world, from asthma and digestive issues to certain types of cancer and neurological ailments like autism and Alzheimer's. All told, chronic illnesses exact an immeasurable toll on individuals and their families in declining quality of life, hardship, and frustration.

The costs are staggering. As of 2016, treating chronic diseases consumed three-quarters of the approximately $3.3 trillion spent annually on health care in the United States, about $5,000 per American. Pharmaceuticals are the treatment for nearly all chronic diseases, and in 2014 American doctors prescribed $374 billion worth of them. That averages out to roughly $1,000 for every man, woman, and child in

America. And though we spend about twice as much on health care per capita than other developed countries do, we—and our kids—land near the bottom of the heap on common health metrics. Even our longevity lags behind peer nations. The sobering reality is that today's children are the first generation in our nation's history expected to live shorter, less healthy lives than their parents. That's not progress.

One of the most significant drivers of chronic health conditions is being overweight or obese for a substantial part of life, as it compromises nearly every major system of the human body. The Centers for Disease Control and Prevention reports that more than two-thirds of Americans are overweight and four out of ten are obese. Our kids aren't faring well either. Since 1970 the number of obese adolescents tripled, and the number of obese children quadrupled. By some estimates obesity-related illnesses now account for more than one in five health care dollars Americans spend.

While the percentage of income spent on food fell by half since the 1950s, the societal cost of health care more than doubled. Our century-long experiment with conventional agriculture seems to have left us better fed than ever in terms of calories consumed but reeling from a slew of chronic diseases. Maybe cheap food isn't so cheap after all.

Our obsession with a calorie-centric view of food has led us to undervalue other components in our diet, especially micronutrients and phytochemicals. Vitamins form parts of enzymes essential to a wide range of normal cell functions. Other micronutrients like zinc support immune system activity and are critical to hundreds of enzymatic reactions in our bodies. Particular phytochemicals activate genes that slow or prevent the onset and progression of chronic diseases. Unfortunately, modern crops and the Western diet shortchange us on these essentials for health.

Micronutrients tied to human health play similar roles in animal health and cows and other ruminants have a strategy for getting the right amount. Given the chance, they'll eat a wide range of different plant forages. Such diversity allows them to select a diet that helps head off or correct nutritional imbalances and deficiencies. And as we'll see, what cows eat ripples through to the nutritional quality of the meat, milk, and cheese we consume.

A century ago the typical Western diet mostly consisted of some meat and minimally processed plant foods grown with few, if any, chemicals. Today it consists of a lot more meat from animals raised in unsavory settings and highly processed plant foods rich in simple sugars and vegetable oils grown with a lot of agrochemicals and synthetic fertilizers. The average American now eats almost five times more calories from added fats and sweeteners than from fruits and vegetables, taking in over 150 pounds of sugar a year, much of it from corn syrup. Although Americans consume almost a quarter more calories a day than in 1961, we're eating far less whole, fresh food, with only one in ten Americans actually eating the recommended daily amount of fruits and vegetables. All in all, more than half of the calories filling up Americans and Canadians now come from ultra-processed foods.

Perhaps we should recast Hippocrates's advice and think of food as *preventive* medicine, as a way to cultivate and safeguard health so as to avoid the need for routine medications in the first place.

Most of us take stock of our health only when we lose it or something threatens it. The same goes for the health or fertility of the land that feeds us. We tend to take our health and that of the soil for granted when good, and when they falter, enterprising folks in medicine and agriculture vie to sell us products to manage symptoms.

Many have heard advice to eat real food, to not eat too much, and to keep physically active. But this leaves out another piece in the puzzle of health. Do the foods we eat contain enough of what our bodies need? Surprisingly little effort has gone toward tracking how farming practices affect the nutritional profile of food. It is generally left to proponents of organic practices to argue that the way we grow food affects our health beyond the obvious point that conventional foods have more pesticide residue.

But there is more to it. The roots of many chronic diseases and autoimmune disorders trace to inflammation, and food choices can either promote or quell chronic inflammation. A diet scant in whole and minimally processed plant foods—like vegetables and whole grains—delivers too few antioxidants and other compounds, including phytochemicals and certain vitamins that help keep inflammation at normal levels. And depending on how farm animals are raised, meat, dairy, and eggs also provide important sources of anti-inflammatory compounds. Conversely, certain foods contribute to inflammation or lack antioxidant power, like refined carbohydrates (sugar), highly processed meats, and many vegetable oils, as well as particular trans fats that do not occur in nature and only recently entered our diet in processed foods.

Changing course is remarkably simple on one level. Overall, we need to eat less but more-nutritious food. This encapsulates the concept of nutrient density, the ratio of nutrients to calories. We all have some idea of what makes for nutrient-dense food. A white-flour cookie or cake loaded with sugar-rich frosting is not nutrient dense. A whole-wheat cracker topped with hummus is. So how do we grow nutrient-dense food packed with phytochemicals and nutrients like essential minerals and fats?

Healthy Soil

Soil health offers a clarifying lens for understanding the effects of farming practices on human health, as it cuts through the usual arguments about conventional versus organic farming. Industrialized organic farms can destroy fertile soil as surely as thoughtful conventional farmers can rebuild the health of their fields. About a third of the world's agricultural land suffers from serious topsoil erosion or reduced fertility that compromises the well-being of at least 3.2 billion people. As if that's not alarming enough, a 2015 United Nations assessment concluded we remain on track to degrade the productive capacity of our planet's agricultural land by about another third over the coming century if farming practices continue to damage the soil. The way we grow our food—what we've been doing to the land—is undermining civilization's agricultural foundation and the health of us all.

On farms around the world, the loss of topsoil and the carbon-rich organic matter on which soil life thrives constitutes an underappreciated global disaster that underscores an odd reality. Today's harvests would dazzle our ancestors. How is it that we can grow more food from worse soil? So far, mineral and synthetic fertilizers, along with the chemicals farmers buy to kill weeds and pests, have compensated for beat-up, ailing soil.

Yet no matter what you may think about industrial agriculture, we cannot simply pin poor soil health on modern farming. After all, Rome's farmers managed to devastate the soils of central Italy, Syria, and Libya long before agrochemicals existed. How? By plowing too much and too often. While they used manure to boost harvests, persistent tillage eroded the fertility of their land right out from under them over the long run.

For millennia farmers around the world have done things to ensure future crops, from praying or offering sacrifices in hope of better harvests to adopting practices that boost fertility. African, Asian, European, and Native American farmers all learned the value of rotating or co-planting different crops and discovered the restorative power of growing legumes like peas, beans, clover, and alfalfa. Over time, traditional farming practices in different regions came to include planting cover crops and letting livestock manure harvested fields. These practices solidified into traditions because they enhanced soil fertility. They worked—at least for a while.

But such practices, especially when done inconsistently or poorly, did not stop water and wind from stripping soil off of bare, plowed fields. Time and again, field after field in region after region lost its fertile topsoil. Yet soil loss usually played out gradually enough to forestall urgency in addressing it.

Widespread soil degradation drove farmers to use—and later wholly rely on—chemical fertilizers as well as pesticides developed for killing insects, weeds, and fungi. These new products increased yields so well that traditional soil husbandry practices that once maintained fertility were neglected, sidelined, or abandoned. A harvest-obsessed world chased progress down the agrochemical highway.

Now, with billions more people to feed, the soil on much of the world's farm- and ranchland stands in far worse shape than under nature's stewardship. With conventional practices continuing to degrade the land, agriculture will change over the coming century—one way or another. The question is how. This unfolding disaster also presents an opportunity to rebuild soil health and realize the great promise of agriculture—harvesting *both* quantity and quality.

Just what is soil health? The U.S. Department of Agriculture (USDA) defines it as the continued capacity of soil to function as a vital living ecosystem that sustains plants, animals, and humans. The idea is rooted in the central role that soil life plays in the processes that cycle nitrogen and other elements from soil into crops, livestock, and people. Indeed, life distinguishes soil from dirt.

Ailing soil leads farmers to depend on measures other than natural soil biology and chemistry to grow abundant crops. The uncomfortable reality is that most of the world's farmland is sick, and increasingly so are we, as farmers prop up harvests with fertilizers and pesticides and consumers depend on pharmaceuticals to combat uniquely modern chronic diseases.

Turning the tide on the historical trend of land degradation will take combining ancient wisdom with modern science to develop and adapt highly productive farming practices that rebuild soil organic matter and restore fertility to degraded fields. Toward this end a regenerative agriculture movement is rapidly growing based, in part, on recognition that farmers harvest higher profits when they restore soil health and spend less on fuel, fertilizer, and pesticides. That healthy soil helps conventional farmers spend less for inputs and still harvest comparable yields produces a buzz at farming conferences across America's heartland, as well as among climate activists. Farmers who boost their soil's organic matter use photosynthesis to pull carbon out of the atmosphere and make it available belowground where it can feed soil life essential to crop health.

Another area of growing interest centers on whether regenerative farmers raise more nutrient-dense food. In asking this question we can look back through history for guidance. Decades ago soil scientist William Albrecht proposed looking at the relationship between soil health and human health as a pyramid. He saw soil health as the foundation for plant health, which supported animal health and ultimately human health. Pioneering organic farming proponent J. I. Rodale laid it out as a simple equation: healthy soil = healthy crops = healthy people. While it remains challenging to connect soil health and human health so directly, we'll break it down along these lines to explore the individual links from soil health to crop health, crop health to livestock health, and both to human health.

Where to begin? Where else but from the ground up, with how life in the soil helps plants eat rocks to build green bodies.

SOIL

ROCKS BECOME YOU

All things come from earth, and all things end by becoming earth.

—XENOPHANES

Imagine walking into a grocery store and scanning the carrots with a handheld device to measure which of them has the most potassium or beta-carotene. Sound like something out of *Star Trek*? Not anymore.

Such a device—the "bionutrient meter"—could kick open the door to greater consumer knowledge, choice, and empowerment. It's the brainchild of Dan Kittredge, founder of the Bionutrient Food Association. The instrument shines a light with a known spectrum of frequencies onto an object and measures what bounces back. The spectral signature is a bit like a fingerprint. The device compares the spectral signature of, for example, a carrot on a store shelf to spectral data in an archival library of carrots with known values of specific nutrients. This real-time assessment would then allow a consumer to pick the best of the bunch. The idea has been used for decades with satellites in mineral exploration, and Kittredge's team cleverly adapted the technology to make a handy device to scan food.

When they began test-driving their gizmo on spinach and carrots, they found a tremendous range in both mineral and phytochemical content. Mineral levels varied by 4- to 18-fold, depending on the mineral. Conservatively, that means a high-quality spinach leaf or carrot could have the nutritional value of at least 4 low-quality spinach leaves or carrots. The variation was even greater in terms of antioxidants and polyphenols important for human health, with the most nutrient-dense carrots and spinach containing up to 200 times more than the least.

But before they can usher in this new era of consumer empowerment, the Bionutrient Food Association needs to build out its database. This will take time, of course. Nonetheless, Kittredge sees consumers eventually armed with the ability to test and compare the nutritional quality of food while shopping as a game changer in reshaping agriculture.

Getting into You

Whatever the specific mineral deficiencies are among carrots or any other crop, they do have one thing in common. For minerals to get into us, they first have to move out of the soil and into the plants and animals that become our food. So how do plants eat? This deceptively simple question took centuries to answer. At first, most of the few who thought about such matters believed plants consumed decaying organic matter in the soil—that plants ate humus. Then in the early 1800s, the discovery that photosynthesis enabled plants to fuel their growth through combining water with carbon dioxide from the atmosphere upended the accepted view. Later experiments established that only part of what makes up humus is soluble and therefore available to plants in the water they draw up through their roots.

These discoveries opened the door for a radically new, chemistry-centric view of the soil that still dominates conventional thinking two centuries later. In short, adding soluble chemicals to the soil proved an efficient and convenient method for getting the nutrients that are key to growth—nitrogen (N), phosphorus (P), and potassium (K)—into plants.

The carbon in plants, people, and other life forms enters the biological realm through photosynthesis. With the exception of a few other elements, notably nitrogen, everything else in living things originally came from rocks or soil as generations of microbes and plant life tapped these sources to build their bodies. Complex molecules essential to all life, like enzymes and proteins, require additional elements, including phosphorus, potassium, calcium, magnesium, and sulfur, to round out plant, animal, and human biomass. Some are abundant in common rocks, others less so. Not coincidentally, along with nitrogen these are the elements commonly found in major mineral fertilizers.

Organic matter, being the remains of once-living things, contains the full range of elements essential to life. So once integrated into biological systems, they circulate through nature's grand cycle of life, death, decay, and rebirth—getting passed around and down through successive generations.

Nitrogen is the geological odd man out among the elements life needs in abundance. It is rare in rocks, yet essential for plants as they need it to make amino acids that assemble into proteins. And protein, of course, makes up a lot of our bodies, including the muscles that move us, antibodies, enzymes, and other molecules fundamental to our biochemistry.

Oddly enough, plants and people alike bathe in nitrogen day in and day out as it makes up almost 80 percent of Earth's atmosphere. But an incredibly stable triple bond holds the two atoms of a gaseous nitrogen molecule together. These atomic twins don't like to separate—or play with other elements. This makes atmospheric nitrogen pretty inert. Plants cannot simply pull it into their leaves through photosynthesis as they do with carbon dioxide. Instead, nitrogen gets into living things through the work of specialized soil-dwelling bacteria that can crack its powerful bond and combine the individual liberated nitrogen atoms with hydrogen or oxygen to make highly soluble ammonia or nitrates.*

Plants also require minor amounts of more than a dozen other mineral elements—like copper, iron, manganese, and zinc. These micronutrients do an enormous number of things in plants. Familiar ones, like iron and zinc, are important for the growth and development of fruits and vegetables. The not-so-familiar tongue twister molybdenum helps plants use nitrogen and turn sunlight into carbohydrates. If the soil is short on mineral elements needed in larger amounts, like calcium or sulfur, or on micronutrients, adding more can boost growth and harvests.

We can't eat rocks, obviously, but we can eat what eats rocks—the green bodies of plants, as well as plant-eating animals. And the mineral micronutrients in their bodies have make-or-break functions in our bodies. Zinc, for example, helps wounds heal and the immune system defeat infections. It is central to the proper working of our senses of taste and smell. Copper is essential for normal heart, lung, and immune system function, as well as cellular respiration and iron metabolism. And iron is key to making the hemoglobin that enables our red blood cells to carry oxygen around our bodies. Consider that the difference between chlorophyll and hemoglobin is one atom out of the 143 composing each molecule. Magnesium anchors the structure of chlorophyll. Iron anchors hemoglobin, and not having enough of it causes anemia. If farming methods limit levels of iron or other elements getting into crops, they affect what's getting into us too.

Mineral elements also play important roles in folding organic molecules like enzymes into different shapes. And all these different-shaped molecules translate into functional versatility that bridges chemistry and biology. In this way mineral elements work like the specialty pieces in a LEGO set. You can only do so much with tiny bricks. Yet it takes

* Lightning strikes are another source of nitrates formed in soils.

just a couple of specialty pieces to put a whole new twist on possible combinations and shapes. Micronutrients open the door to making enzymes that direct cellular maintenance and are critical for health, even though we only need small amounts of them relative to the other elements that build our tissues and fuel our cells.

Phosphorus is a particularly important element that plants and animals need. It is the second most abundant mineral element in your body, a key part of DNA and cell membranes, critical for building bones and teeth, and a must-have for how our bodies use and store energy. But it is naturally quite scarce in most rocks—and therefore in mineral soil. Most unhelpfully for plants, phosphorus in the soil readily oxidizes to form stable, rather insoluble compounds. So what little is present generally sits parked in the geological realm, locked out of biological circulation. Yet just as for nitrogen there is another source of phosphorus and other mineral elements critical to life—dead things, organic matter. It just takes microbial partners to mine and deliver it.

Tiny Partners

Nature long ago figured out how to recycle vital mineral elements back into living things. The soil is where this happens. Consider that nobody fertilized tropical jungles, the native prairies of the Great Plains, or old-growth rainforests of the Pacific Northwest. Yet the lush growth of plants in all these environments relies on getting both major and minor elements out of the soil and into their bodies through roots that only reach so far. How did they manage? Trillions of tiny allies.

Soil-dwelling bacteria and fungi are the worker bees of the subterranean world. They produce enzymes and organic acids that release the minerals held in rocks and organic matter for plants to take up. And when soil fungi and bacteria consume organic matter, they and the creatures that in turn eat them metabolize and convert their meals into soluble nutrients that plants can take back up and use again. In this way, soil microbes act as a procurement and supply system for plants, particularly for less commonly mobile mineral elements like phosphorus, iron, and zinc.

Ingesting and digesting food breaks it down to molecular building blocks our bodies reassemble into our cells and tissues. At one level, that's the whole biological point of eating. We get quite a bit of help from our gut microbiome. In contrast, land plants handle digestion outside of their bodies, relying on the soil microbiome to serve as an external stomach, prepping as well as supplying many of the nutrients and compounds plants need to build their bodies.

And the minerals that soil microbes help make available to plants set what gets into our crops and the animals that, in turn, eat them. Consider zinc. Different soils will have different amounts, but none will have a lot. Mycorrhizal fungi serve as root extensions for plants, locating and extracting it from soil particles and organic matter to hand over to plants. A single shovelful of soil can contain miles of thinner-than-a-thread fungal hyphae, making for a vast underground nutrient mining and transportation system. Not all soil-dwelling fungi attach themselves directly to plant roots. Nonmycorrhizal types specialize in decomposing organic matter, which releases nutrients in forms plants can readily take up.

Many crops, including wheat, corn, rice, soybeans, potatoes, bananas, yams, flax, and coffee, form partnerships with mycorrhizal fungi. Why would fungi help plants take up minerals like zinc? Fungi themselves can't photosynthesize, so they make trades with plants—zinc for sun-made foods. Such exchanges occur in the wild frontier beneath our feet in a place called the rhizosphere (from the Greek *rhiza* meaning "root"). This halolike zone extends from a few millimeters to a centimeter or so around each and every root and root hair. It's a lively place. The soil immediately around plant roots hosts a grand and busy bazaar with nutritious meals for microbes exuding from the roots of a plant into the rhizosphere. Mycorrhizal fungi and soil-dwelling bacteria lap up these exudates and in return give plants nutrients and metabolites that boost botanical growth and defense. Mycorrhizal fungi also convey underground signals to plants that help tee up defenses to repel pathogens like those that cause tomato blight. Some fungi even help plants share nutrients. The benefits plants receive for stocking the rhizosphere with exudates underpins the botanical world's health care plan.

For robust growth and health, both plants and people need a full slate of macro- and micronutrients. Commercial fertilizers, however, tend to supply only macronutrients, like the standard trio of nitrogen, phosphorus, and potassium; calcium and sulfur in gypsum; or calcium and magnesium in pulverized limestone (chalk). Such mineral and synthetic fertilizers do spur crop growth—a lot, especially in degraded soils. Nitrogen fertilizers in particular boost yield and protein content. But as the agronomic trinity of mechanized tillage, agrochemicals, and monocultures radically altered how we grew food, it disrupted connections central to how soil life acquires and delivers mineral elements to plants.

It stands to reason that disrupting ancient relationships in the rhizosphere profoundly influences the movement of micronutrients into crops. This in turn affects what gets passed on to livestock and into us. If nutrients are low or lacking in crops, we and other animals get shorted too. The value of soil microbes for soil health is now coming into sharper focus for agronomists long trained to prioritize chemical solutions over fostering long-running symbiotic relationships between crops and soil life.

People considered fungal mycorrhizae an inconsequential curiosity when first discovered. But by the 1930s the importance of soil life to plants started coming into focus. One of the first academic researchers to recognize the botanical importance of microbes was Selman Waksman, a Rutgers University professor who coined the term *antibiotic* after discovering that a soil bacterium could produce what became the wonder drug streptomycin. In their 1931 book *The Soil and the Microbe* Waksman and coauthor Robert Starkey described the key role soil microbes play in decomposing once-living matter, a process that keeps elements circulating from soil to plants and animals and back again. Although Waksman began to appreciate what soil bacteria and fungi did for plants, *why* microbes did it remained cryptic. It was only decades later that deciphering the chemical language between plants and microbes unveiled complex interactions central to negotiating the challenges of the botanical world's stuck-in-place lifestyle.

Understanding reciprocity in an agricultural context means understanding that carbon is the main currency in nature's underground economy. It makes up about half of soil organic matter. The exudates that flow out of a plant's roots into the rhizosphere are another source of carbon. Since soil as a whole is carbon-poor relative to the rhizosphere, most soil microbes live in or very close to the rhizosphere. The specific organic compounds and mixtures that plants release into the soil—sugars, proteins, organic acids, vitamins, enzymes, and even fats—attract specific microbes to colonize the rhizosphere. These recruits are a diverse lot. Some physically block pathogen access to roots. Others produce antimicrobials that drive away specific pathogens. In so doing, plants actively recruit and support communities of microbes instrumental to their own well-being, establishing relationships that lie at the heart of plant—and crop—health.

Dead plant tissues like leaves, stems, and roots that make up organic matter are another part of the underground economy, as are dead bacteria, fungi, and other microbes. But conventional farming practices, especially tillage, cause soil organic matter to break down at a rapid pace. A 1994 study in *Nature* reported that fields in tropical regions could only be tilled economically for less than a decade. After that, so little organic matter remains that supplemental fertilization is required. Likewise, the natural productivity of temperate prairie soils typically lasts just over half a century under regular tillage.

When the rate at which carbon is released from soil organic matter races ahead of the ability of soil life to reincorporate it into more biomass, the excess carbon doesn't just disappear. It heads skyward. A late 1970s study found a third of the carbon added to the atmosphere since the industrial revolution came from degrading organic matter in agricultural soils. So far, North America's farmland has lost about half of its natural endowment of soil organic matter. Roughly the same goes for agricultural soils globally. As organic matter and the soil life it supports are key to maintaining fertility over the long run, the acute loss of so much from soils around the world threatens to unravel a foundational thread running through the tapestry of life.

The complex exchanges and interactions occurring belowground between plants and soil life are as intricate and finely wrought as the relationships between plants and pollinators familiar to us aboveground. Historically, biologists tended to emphasize competition, embodied in Tennyson's famous line about "nature red in tooth and claw." But more recent discoveries in microbiome science and ecology have led to new realizations. Symbioses are far more common and widespread than once thought. In this light the rhizosphere shines as a dynamic living shield that protects and provides for all involved.

Obviously, microbes don't have intentions—good or bad. They are brainless, single-celled organisms after all. But they can be good in the sense that certain microbial communities are essential to the health of plants and people. It turns out that the vast majority of microbes either do us no harm or help us in some way. But "good" microbes, like opportunistic political allies, will only hold up their end of a relationship as long as interests align. They can turn on their host if circumstances change, as with a plant that fails to hold up its end of the deal to supply exudates. When we repeatedly create an ecological blank slate through routine, indiscriminate use of broad-spectrum biocides in agriculture, we knock back a complex microbial ecosystem that provides an awful lot of benefits. The same goes for overuse of antibiotics in medicine.

And whether in the soil or the human gut, weedy species rebound first when given the chance to repopulate our fields and bodies.

Where has our century-long war on microbes landed us? We tamed many infectious diseases but now face superbugs and a modern plague of autoimmune and chronic diseases. On farms, parallel developments led to increasing reliance on pesticides to counter pest outbreaks arising from flagging soil fertility. All that progress left us hooked on pharmaceuticals and agrochemicals.

What would a different frontline strategy look like, one that could preserve the effectiveness of antibiotics and pesticides for when we truly need them? One option is cultivating the microbial allies that benefit us when we partner with them.

We can look to evolutionary history for perspective on why fungi and plants form partnerships. When plants first came ashore more than 400 million years ago, their roots served mostly as an anchor while fungi provided them with minerals from the soil. How do we know? Early fossil land plants have mycorrhizal fungi entwined with their primitive roots.

While fertilizers have kept crop yields high enough for us to overlook the problem of declining soil organic matter, we now understand the consequences. Starving soil microbial communities of sustenance restricts their ability to deliver key micronutrients and other beneficial compounds to crops. We traded away quality in pursuit of quantity as modernized farming chased higher yields, overlooking a farmer's natural allies in the soil. As we'll see, mineral loss is not the whole story on nutritional quality, but so far it has received the most attention.

Losing Ground

Among the factors suspected or generally blamed (incorrectly, as we'll see) for declining mineral levels in crops is depletion of minerals from the soil itself. Far less controversial is the effect of crop breeding in trading off nutrient content for high yields. Though it is not widely acknowledged beyond agricultural circles, this well-known reality has a name among agronomists and soil scientists—the dilution effect.

Consider what happens when we breed crops for larger edible parts—bigger roots, fruits, or shoots. Picture a small fingerling potato next to a hefty russet, both grown in the same soil and that take up the same amount of iron. You'd be eating a different amount of iron in identical serving sizes of the two spuds. The smaller fingerling would contain more iron per bite than the russet, making the former more nutrient dense.

Another case of the dilution effect occurs in wheat bred for higher yield. More seeds per seed head mean minerals are spread thinner. This is troubling as grains are both staples and particularly good sources for mineral micronutrients in the human diet.

USDA agronomists who analyzed wheat varieties introduced from 1873 to 1995 found significant declines in iron, zinc, and selenium after conducting side-by-side field trials at two locations in Kansas. They concluded that crop breeding for higher yields reduced mineral micronutrient uptake and that soil conditions also played a role. A similar study from Washington State University compared 63 historical and modern wheat varieties and found that higher-yielding modern types had significantly lower concentrations of copper, iron, magnesium, manganese, selenium, and zinc.

Researchers in England came to a similar conclusion based on archived wheat samples from the Broadbalk Wheat Experiment, the world's longest-running agricultural field trial. They found that the concentrations of copper, iron, magnesium, and zinc remained stable from 1845 to about 1960, and then decreased significantly through 2005. They also measured the amount of mineral micronutrients in the soil but found no evidence of depletion in the soil itself. They therefore concluded that the Green Revolution unintentionally decreased mineral density in wheat due to the introduction of high-yielding crop varieties.

Reduced mineral content is not unique to wheat. An analysis of the changes in mineral composition of British fruits and vegetables from 1936 to the mid-1980s found statistically significant declines in calcium, magnesium, and copper in vegetables and magnesium, iron, copper, and potassium in fruits. Phosphorus was the only mineral element that reportedly did not decrease over this period.

Likewise, a 2004 comparison of published USDA food composition data from 1950 and 1999 across a group of 43 garden crops (39 vegetables, 3 melons, and strawberries) found statistically robust declines in protein, calcium, phosphorus, and iron, as well as in vitamin B_2 (riboflavin) and vitamin C (ascorbic acid). While the study found substantial variability within each individual food, the mineral content of some foods actually increased. This was a clue that nutrient losses were not simply due to broad depletion of minerals in the soil.

A subsequent review of food survey data from the 1930s and 1980s in the United States and the United Kingdom found major decreases in the average concentrations of copper, iron, and potassium in fruits. In vegetables, the average concentrations of copper and magnesium decreased in Britain, as did average concentrations of calcium, copper, and iron in the United States.

A 2009 University of Texas study that reviewed previous studies of declining nutrient levels in U.S. crops concluded that there was strong evidence for 5 to 40 percent declines in the mineral content of fruits and vegetables over the previous half century. This study also noted that reported values of calcium in more than two dozen commercial broccoli varieties were substantially lower than published USDA reference values, which themselves decreased by about two-thirds from 1950 to 2003. As the nutrition we expected our food to hold ratcheted down, we got less from growing more.

While depletion of minerals in the soil is often identified as a likely culprit for declines in crops, a 2017 review of prior studies on nutrient declines found little evidence to support this interpretation. Instead, the review attributed the observed decline in crop mineral density to the introduction of new crop varieties. In the end, however, the author emphasized the benefits of higher yields on the global food supply and dismissed the importance of nutrient dilution effects, arguing that they could be offset through eating more fruits, vegetables, and whole grains. The message, apparently, was that nutrient depletion was simply the price to pay for high yields.

But look at the problem of declining mineral levels another way and perhaps the answer lies in moving minerals from soil into crops rather than consuming more food that is less nutritious.

A Simple Experiment

Can increasing levels of soil organic matter and rebuilding populations of microbes that act as nature's miners and truckers change mineral levels in crops? We had an opportunity to find out after presenting a public talk in southern Washington State when a soil conservationist with the Natural Resource Conservation Service relayed a natural experiment taking place on a farm just across the border in northern Oregon.

A wheat farmer wanted to find out what would happen to yields on two adjacent fields under different methods of weed control. One method was the region's conventional practices (bare fields rotated with winter wheat and regular applications of the herbicide glyphosate). The other was a new practice for him—a complex crop rotation of spring barley, winter wheat, and a diverse mix of cover crops. The idea was to compare whether keeping the fields vegetated year-round could outcompete weeds without using herbicides. The farmer planted both with the same variety of wheat. He then fertilized with compost and a "starter blend" of chemical fertilizers. After two years he had an answer to his yield question: both fields produced a harvest of 75 bushels an acre. The cover crops suppressed weeds without reducing his harvest. But what about the mineral levels in the wheat?

The farmer sent us samples to analyze, and the results intrigued us. He had added no supplemental mineral micronutrients to the no-till, no-herbicide, cover-cropped field. And yet, relative to the wheat grown under conventional practices, the wheat from the cover-cropped field had 35 to 56 percent more boron, manganese, and zinc and 18 to 29 percent more copper, iron, and magnesium.

These increases in mineral micronutrients were comparable in magnitude to reported historical nutrient declines. Yet you can't change the bulk chemistry of the soil in just a couple of years. Something else was going on, something important.

What could change that fast? Soil biology, as it turns out. It seems that in planting cover crops and forgoing herbicides the wheat farmer grew legions of underground microbes that ferried minerals from the earth into his harvest.

LIVING SOIL

Healthy citizens are the greatest asset any country can have.

—WINSTON CHURCHILL

Suspicions about the influence of farming practices and soil health on the nutrient density of foods and human health date back almost a century. On March 22, 1939, a panel of 31 English physicians gathered at the Crewe Theater in Cheshire to discuss the medical evidence and connections they saw between diet, farming, and health. The audience of hundreds included family doctors, farmers, politicians, and interested members of the rural public. The panel assessed the state of British medicine based on their own patient experiences and consultations with the region's family doctors. The evidence, laid out in a pamphlet supplemented with 28 pages of references, led them to propose that the accelerating use of chemical fertilizers and pesticides damaged human health.

As he opened the meeting, the panel's chair praised the British health care system's progress in extending life expectancy. He noted how this mostly reflected greater public access to the services of his fellow physicians. But, he went on, the nationwide "fall in fatality" was "all the more notable in view of the rise in sickness." The English were living longer, yet less healthy lives and increasingly seeking treatment for readily preventable chronic maladies. What culprit did these doctors blame for undermining the 25-year-old National Health Insurance Act's goal of preventing and curing sickness and disease? The English diet.

Their indictment centered on four well-known and prevalent English health conditions: bad teeth, rickets, anemia, and digestive disorders. In each case, the panel pointed to evidence of poor diet as the root cause. It wasn't that people lacked enough to eat. Their food lacked adequate nutrition. The panel cited a 1936 study of millions of English schoolchildren that found more than two-thirds had major dental problems. Genetic disposition could not explain the contrast to the nearly perfect teeth of the English population of Tristan da Cunha, an outpost of the British Empire in the South Atlantic. The panel also decried the annual scourge of thousands of cases of bone-softening rickets among English schoolchildren, despite "its prevention by right feeding" being "so easy that every dog breeder knows the means."* Anemia ravaged poor children, and digestive disorders afflicted a sizable and growing proportion of the populace. The doctors ascribed these and many other ailments to nutritional deficiencies. They saw no way around it. The modern English diet was not healthy.

The panel concluded that British doctors spent half their time on health conditions stemming from poor nutrition. They thought reducing human illness would require improving food quality, which in turn would come from restoring soil fertility.

Confident as they were in what they were seeing, the doctors were less certain as to exactly what was missing from modern food. They pointedly did not advocate any particular diet, citing how healthy Esqui-

---

* Vitamin D is particularly important for preventing or treating rickets as it enhances absorption of calcium and phosphorus, the two main components of bone. Eggs and fatty fish are good sources of vitamin D, and some foods (e.g., dairy products) are now fortified with it.

maux (as they called the Inuit) lived on "flesh, liver, blubber, and fish"; the Hunza and Sikhs in India lived well on "wheaten chappattis, fruit, milk, sprouted legumes, and a little meat"; and the British inhabitants of Tristan da Cunha thrived on "potatoes, seabirds' eggs, fish, and cabbage." In laying out their case, the doctors pointed to the common element among the diets of remarkably healthy peoples—a preponderance of fresh, whole, and minimally processed foods.

Something was missing, they claimed, from England's working-class diet of white bread and boiled vegetables grown with chemical fertilizers. Drawing on veterinary medicine, they concluded that new Western diets and farming practices translated into nutritional deficiencies as surely in people as livestock. They considered it their duty "to point out that much, perhaps most, of this sickness is preventable and would be prevented by the right feeding of our people."

After the committee chair finished presenting the panel's findings, two prominent voices knighted for their work on food and agriculture backed up the doctors' testimony. Sir Robert McCarrison addressed what to eat, and Sir Albert Howard addressed how to grow it. After the presentations, the panel put the question of whether to endorse their conclusions to the audience. With little debate, the attendees unanimously carried a motion to accept the panel's assessment. Being English, the crowd then adjourned to the adjacent town hall for tea.

McCarrison had spelled out what he saw as the big problems with the English diet—eating lots of processed flour (simple carbohydrates) and meat coupled with too few fresh vegetables. He also maintained that dairy intake was too low. Having performed groundbreaking experiments showing that diet and food quality influenced health, McCarrison set a scientific foundation for the panel's perspective.

In the opening days of the twentieth century McCarrison served as a medical officer along the rugged Himalayan frontier in colonial India. The experience provided him with ample fodder for his later ideas about diet and health. In particular, he noticed striking differences in the physique and health of the region's inhabitants. Suspecting diet as the cause, he decided to experiment on rats, which eat pretty much anything people eat. McCarrison considered a year in the life of a rat akin to decades of a person's life. In just a few months' time he thought he could glean meaningful insights into the lifelong effects of diet on human health.

His rats thrived when he fed them the same foods people "whose physique was good" ate—whole-wheat bread, milk, fresh vegetables, and pulses (the edible seeds of legumes like lentils, peas, and beans). For two years he raised a group of rats on this diet and found they suffered no illness or death from natural causes. Nor did they experience infant or maternal mortality. In stark contrast, another group of rats ate a diet with a minimum of milk and fresh vegetables, patterned after that of people "whose physique was bad." These rats developed stomach, bowel, kidney, and bladder diseases mirroring those of their human counterparts.

Intrigued, McCarrison kept experimenting. He divided a group of juvenile rats into large cages where he kept everything the same except for their food. Rats in one group received a "good" diet consisting of whole grains, milk, fresh vegetables, and occasional fresh meat. Those in the other group ate the typical English diet of "white bread and margarine, tinned meat, vegetables boiled with soda, cheap tinned jam, tea, sugar and a little milk." The difference proved striking.

Rats raised on the "good" diet enjoyed physical and social health. Neither disease nor conflict plagued them. Those fed a typical English diet did not fare so well. Many became ill and began to fight. After two months they started to kill and eat one another. McCarrison closed his remarks by noting that rats fed poor diets developed many of the same diseases that increasingly afflicted the English. To him, the lesson was clear. Diet mattered to health, and eating fresh, whole foods was the diet that delivered.

Sir Albert Howard, who also reaped enlightening insights working in India, spoke next. He argued that fertile soil supported public health, producing healthy food that served as preventive medicine. After arriving in India in 1905 to serve as Imperial Economic Botanist, Howard saw how farming practices influenced crop health and the nutritional value of food. He noticed that local subsistence farmers had no trouble with the pests and diseases that plagued the owners of large tea estates he was there to advise.

Fascinated by these differences, he studied local composting methods, eventually adapting them for use on tea estates. The return of organic matter to the land lay at the heart of his scaled-up system. Time and again, Howard found that the soil on estates and subsistence farms treated with "properly made compost" led to disease-resistant crops and that livestock fed on these crops proved healthier and more resilient than those fed on chemically fertilized crops. He considered the health of soil, plants, animals, and people an interconnected chain.

Howard captivated the Crewe Theater audience with his story of learning about the benefits of compost that modern agronomists overlooked. He told of insights gleaned from the peasant farmers of India, whom he called his best teachers. He considered the insects and fungal pests that attacked weakened crops nature's professors of agriculture. They helped him recognize undernourished and diseased plants as symptoms of perturbed soil ecology.

Howard concluded that the key to growing healthy, nutritious crops was to nourish soil life, nature's recyclers. And to him the way to do that was clear: provide them with high-quality nutrition in the form of properly made compost as well as animal wastes like manure and nitrogen-rich urine. Soil fungi and other life would eventually deliver minerals and other nutrients in the compost back to plants. But this did not happen instantaneously. Only when fungi and bacteria died could the elements in their bodies return to the soil as food for plants. Howard's point to the audience was that a well-made compost or mulch fed and supported beneficial soil life. Chemical fertilizers did not. McCarrison and Howard were not exactly radical anti-establishment types. These titled luminaries built their ideas from decades of research and experiments focused on the role of diet in preventing illness in crops and animals.

A month after the public meeting, the *British Medical Journal* published the panel's pamphlet as "Nutrition, Soil Fertility, and the National Health." It triggered heated commentary in the normally staid periodical. Not everyone, it turned out, shared the enthusiasm of those at the meeting.

One physician in particular, New Yorker Richard Bomford, published a letter in the journal that channeled the response of mainstream academics and doctors. After concurring that "no one would question the importance of good nutrition and . . . diet on physique and health," Bomford lambasted the view that food grown using chemical fertilizers was less nutritious. "Let us not abandon scientific method for mysticism," he implored. Growing vegetables using "inorganic salts" (as fertilizers were called at the time) had become commonplace. He saw it as incumbent upon those criticizing modern methods to demonstrate how growing food with chemical fertilizers produced nutritional shortcomings. If everyone was doing it, how could it be so bad?

Looking back, we can see in Bomford's critique a foreshadowing of Monsanto's infamous ad campaign of the 1970s, "Without chemicals, life itself would be impossible," and DuPont's ingenious "Better living through chemistry" slogan. Across decades, the implication remained the same. Scientists, doctors, and the public alike should not question advances in industrial agrochemistry. (It also heralded the now-familiar pattern of sowing doubt to impugn, belittle, and delegitimize scientists studying the effects of pesticides, tobacco, and climate change.)

In response to Bomford's skeptical missive, the journal published a letter from Howard pointing to the 28 pages of references accompanying the Cheshire doctors' pamphlet. Their medical testament, Howard wrote, was well grounded in evidence—in rational thought, not starry-eyed mysticism. He went on to school Bomford that if one bothered to look, it was easy to see the health effects of differences in nutrition when using "artificial manures," as Howard called chemical fertilizers. He pointed to examples of how a change in diet from chemically fertilized crops to ones grown from compost-manured soil dramatically improved the health both of animals on a farm in Surrey and of students at a large boarding school near London.

The explanation as Howard saw it was radically simple. Chemical fertilizers undercut the work of soil fungi and bacteria, compromising their delivery of micronutrients critical for plant nutrition and thus the health of crops and livestock—and people. He intuited that mycorrhizal fungi partnered with plant roots in some way, although he and others at the time didn't fully understand how.

Howard maintained that a few chemical fertilizers could not do the same things as soil microorganisms. This was not a rejection of modern science, as his critics were quick to cast it, but recognition of an expanding frontier of discovery where insights based on observations and experience raced ahead of mechanistic understanding.

His work had shown that crops raised on humus-rich soils were better able to resist disease and conferred health and resilience to livestock that consumed such crops. What was the mysterious something that imbued food grown on compost-fertilized fields with nutrition vital to health that food grown with chemical fertilizers lacked? He faced the same problem as the doctors—demonstrating how the connection worked. And without being able to explain exactly how soil life acquired and passed on nutrients, Howard's peers considered his views more spiritualistic nonsense than modern science. It took half a century for science to catch up and fill in gaps in our knowledge of plant and soil science, as well as microbiomes. In the end Howard may not have fully understood how it all worked, but he was on the right track.

The Green Queen

A quarter century younger than Sir Albert Howard, Lady Eve Balfour further connected agricultural practices and public health, putting her too at odds with most scientists of her day. That her 1943 book *The Living Soil* remains in print and relevant today testifies to the groundbreaking nature of her ideas.

Evelyn Barbara Balfour was born in 1898, one of six children of the Earl of Balfour. The niece of a former prime minister, Eve was no stranger to upper-class society. But she preferred the rural life of a country farmer. During the First World War, she studied at Reading University College and became one of the first women to receive a degree in agriculture from an English university. After the war ended, she and her sister used their inheritance to buy a farm in eastern England.

For two decades they raised crops, dairy cattle, sheep, and sometimes pigs. In 1938, Balfour absorbed Howard's ideas about compost-based farming and the role of soil life in supporting healthy crops, livestock, and people. She began experimenting on her farm and rapidly became convinced of the need to expand the conventional chemistry-based view of soil fertility to include soil biology.

In the run-up to and during the Second World War, shortages, want, and loss across Britain drove her thinking on a living soil. She built her case around her own experiences and drew on the work of Howard and a network of medical doctors. Weaving these threads together she argued that soils filled with life set the stage for healthy plants, animals, and people.

Reading her book today, one is struck by Balfour's affinity for ideas at the forefront of thinking in the modern soil health movement. She pieced together evidence, observations, and experiences from farmers, agronomists, and medical doctors that challenged conventional views of farming and soil fertility, arguing that soil ecology was the key to fertile soil and healthy crops. Balfour maintained that a diet of fresh, whole foods grown in healthy soil was a recipe for healthy people. She even went so far as to call for hospitals to retain soil scientists on staff.

Like Howard, Balfour thought mycorrhizal fungi and soil bacteria provided nutrients, aided in maintaining plant health, and contributed to nutrient-rich food. She marshaled indirect evidence, drawing together descriptive studies and experiments that suggested connections. But it took decades for advances in soil microbiology to confirm links and fill in mechanisms for the favorable effects she saw play out in practice. She too could see what worked but couldn't demonstrate how. This did not impress a skeptical and increasingly specialized scientific community that dismissed her ideas with patronizing relish.

Neither deluded nor discouraged, Balfour presented the evidence she had gathered to make her case, noting the transformative implications of establishing that soil health translated into human health.

> If, when it is concluded . . . beyond any reasonable doubt that health is in fact to a large degree dependent upon correct soil management, then we should be faced with a revolutionary situation, for clearly in that event, any Public Health system of the future would have to be based on soil fertility.*

Arguing that the health of human societies depends on the health of their land, she maintained that the consequences of ignoring this fundamental connection played out tragically in the past. Relying on intensive fertilizer and pesticide use presented a new and mostly unrecognized threat to the soil and human health.

* Balfour, E. B. 1943. *The Living Soil: Evidence of the Importance to Human Health of Soil Vitality*. Faber and Faber, London, p. 24.

In her view, starving beneficial soil life opened the door for pests and pathogens—and spurred the growing demand for pesticides.

> Parasites and diseases appeared in the crops, and epidemics became rife among our livestock, so that poison sprays and sera had to be introduced to control these conditions.*

But just what was the connection between soil health, plant health, and human health? Balfour suspected that soil biology played a bigger and more central role in plant nutrition and health than generally recognized. Soil life, it seemed, was consuming organic matter and transforming it into other things important to plant health. Unlike Howard, Balfour correctly concluded that it was not that plants consumed fungi to get nutrients but that fungi made nutrients available to plants.

In her opinion academic research fixated on the causes of disease and didn't focus nearly enough on another question—the source of health. Balfour saw that crops suffered far less disease if grown in fertile soils with adequate organic matter. And animals that were fed on or foraged from such plants also exhibited better disease resistance. These effects, she asserted, would ripple through to people who consumed such plants and animals. She saw this causal chain as the key to improving the health of England's sickly working class.

These were a lot of dots to connect. Yet Balfour meticulously laid out her case, drawing on examples from her farm as well as published studies and the experience of others. But she and like-minded mavericks still lacked solid evidence for how soil-dwelling fungi secured crop health. Without the explanatory power of identified mechanisms, her ideas languished in scientific circles. So she decided to turn her farm into a grand experiment.

* Balfour. *The Living Soil*. p. 51.

The Haughley Experiment

Balfour's motivations sprang partly from well-authenticated observations that changing from a diet of highly processed food to fresh, whole foods improved health (and behavior). Confident that agricultural researchers would not invest the time, money, and effort in testing what they saw as yesterday's practices, Balfour decided to tackle the problem of declining soil health herself.

The strongest evidence for the role of soil conditions on health and nutrition came from the botanical world. Balfour's correspondents commonly reported less plant disease and greater resistance to pest damage under organic practices than under chemical fertilization. Studies at the time had also shown that organic farming practices improved livestock health.

Her experience ran counter to orthodox thinking and bridged fields in unconventional ways. Medical research tended to focus on causative factors for specific diseases rather than the ability of diet to maintain good health. The Ministry of Agriculture considered such studies to be a matter for the Medical Research Council. They, in turn, considered them the purview of the Ministry of Agriculture. Unimpressed with the myopic research both groups funded, Balfour came up with a different idea: to compare the effects of food grown using organic and conventional practices on animals, and to continue the experiment not for just one season but over successive generations of crops grown in adjacent fields.

The opportunity to set up the experiment no one else would do came in 1939 when Balfour's neighbor, Alice Debenham, left her 80-acre Walnut Tree Farm to a trust established to conduct research on the relationship between humus and health. Balfour added 136 acres of her own farm to create the Haughley Research Farm. She subdivided the combined property into three units. An organically farmed section was to receive only crop residues and animal manures produced on that section. It was to emulate a closed system as closely as possible. A second section was to receive chemical fertilizers in addition to the crop residues and manure produced on it. The third section would have no livestock, and crops would receive only chemical fertilizers. In the two sections with livestock the animals would eat only crops and forages that grew in their respective section.

Three sections were up and running when the Second World War interrupted plans to raise support for research. Nonetheless, Balfour managed to obtain an exemption from wartime orders regulating chemical fertilizer use so she could maintain distinct practices on each section of her new research farm.

Publication of *The Living Soil* in 1943 raised interest in the experiment, but fundraising proved difficult during and after the postwar period. Society was rebuilding and looking forward, not investing in new research on old ideas. But Balfour and peers persevered, and in 1946 she cofounded the Soil Association. The work at Haughley Research Farm carried on and expanded to promote what we now call sustainable agriculture.

The experiment continued through two decade-long rotations that included oats, barley, pulses, wheat, and a multispecies pasture mix that was grazed for four years before starting back with oats. Balfour insisted comparisons be made over this full crop rotation. By 1952 the first rotation began after the founding livestock had been purchased—Guernsey cattle, light Sussex poultry, and homebred pigs (which were replaced by sheep in 1960). The section without livestock followed a five-year rotation that would run through twice during a full rotation on the stocked sections.

After the first complete rotation ended in 1961, the Soil Association reported the results to members. They found slightly higher yields from the two fields where chemical fertilizers were used but noted that yields on the organic section did not decline over two decades. Their results showed that fertility could be maintained without the addition of chemical fertilizers. Indeed, the highest-yielding portions of the organic section were those that had gone the longest without chemical fertilizers. And while the yield of the organic cereals was generally lower, their protein content was consistently higher. In addition, dairy cows grazed on the organic section proved healthier and produced 15 percent more milk.

Notably, the soil changed over the course of the experiment. By 1960, the humus content of the organic fields rose to about 6 percent, whereas that of fields with no livestock fell to 3 percent. The experiment also demonstrated that the organic soil could hold half again as much water as the fields without animals. The soil in the conventionally managed stockless section made for dusty, hard, crusty ground that lay submerged under ponded water after light rainfall. In contrast, even heavy rains soaked into the fluffy, spongy soil of the organic fields.

The amount of humus in the organic section was consistently higher than in the mixed section that received both manure and chemical fertilizers. Contrary to conventional expectations, chemical fertilizers were not boosting soil organic matter.

And there was another finding that challenged mainstream thinking: despite the lack of chemical fertilizers, plant-available nitrogen in the soil was highest on the organic section. In contrast, the stockless section did not increase in soil nitrogen, even though it received annual additions of nitrogen fertilizer. Balfour concluded that routine fertilizer applications proved wasteful. Crops took up less than half of what went onto the fields, with the rest running off to pollute downstream waters.

Phosphorus also behaved differently in the organic and conventional fields. On the stockless, conventionally fertilized section, levels of soluble phosphate in the soil proved highly variable, with a noticeable peak in late fall after fertilizer application. The plant-available phosphate levels in the organic sections, however, peaked in late summer, a pattern attributed to enhanced biological activity in the soil, with the highest levels measured in the fields with the highest humus content.

This finding in particular led Balfour to a realization. Soil organic matter enhanced microbial activity. Here was a connection between how farming practices influenced soil life and deliveries of mineral elements to plants and crops, which then rippled through to livestock and people.

Balfour considered the findings related to sources and seasonal fluctuations in plant-available minerals a major contribution of the Haughley experiment. It meant that soil life made minerals available to plants in the amounts required when crops needed them. That soil life was a key ingredient modulating fertility flew in the face of orthodox thinking.

In addition, Haughley Farm's organic crops exhibited greater resistance to damage from insect pests. They rarely showed any symptoms of mineral deficiency. Weeds increased on the mixed and stockless sections where herbicides were regularly used—so much so that spraying became essential to maintain crop production. Far less weed pressure arose in the organic section. In 1960 and 1961 weevils threatened crops on the mixed section, while crops in the organic section remained unscathed. And in 1961, the oat-and-pea mixture planted for silage on the mixed section was lost to weevils. No apparent damage occurred to the same crop on the neighboring organic section.

Another major finding was that the organic section produced healthier animals. The herd of cattle in the organic field was consistently in better condition and exhibited "greater contentment and placidity" than those on the mixed section. They also produced more milk per unit of feed despite eating less overall than those in the mixed-section herd. The Haughley report attributed the finding that total milk production per acre was greatest on the organic section to the total solids content of forage in the organic pasture being almost twice that of the mixed pasture. In other words, cows on the mixed pasture had to eat twice as much grass to get the same amount of nutrition. In 1963, chemical fertilizers were intentionally omitted from a strip running across the center of the mixed field. When cows were turned out to graze, they gobbled up the unfertilized strip before grazing the rest of the field.

On average, the cows in the organic field produced milk for several more years than the mixed-section cows. And adult mortality among the chickens was lower in the organic flock. Regional farmers who purchased Haughley chickens consistently reported that birds from the organic section proved more productive and resistant to disease than those from other flocks.

In 1961, the amount of plant-available trace elements in the soil was compared. Relative to the stockless (conventional) section, the organic section held 6 to 20 percent more plant-available manganese, molybdenum, zinc, copper, boron, cobalt, and iron. Magnesium was the same for both sections, and the stockless section had a quarter more nickel. In both 1962 and 1963, plant-available levels of iron, magnesium, and zinc were higher in the soil of the organic section. While the micronutrients in harvested crops proved too difficult for the team to measure, another nutritional result stood out. The vitamin C content of the milk from the herd raised on the organic field was 15 percent higher than the milk from the herd raised in the field receiving chemical fertilizers.

Another interesting finding came from watching the behavior of the herd in the organic section. They grazed two adjacent pastures with different mixes of forage. One pasture consisted of a wide range of deep-rooted herbs, and the other, grasses and clovers. The revealing behavior happened when cattle were given access to dried seaweed in both pastures and allowed to move from one pasture to the other. Whenever the cows grazed on the deep-rooted herbs, they didn't bother to eat the seaweed. But when they grazed on grasses and clovers, they quickly consumed all the available seaweed. Across the whole farm, cattle fed most avidly on deep-rooted plants. Clearly, these cows were not mindlessly grazing. They could detect differences among forage plants and act on their preferences.

The Haughley experiment ran until 1969, when it was discontinued due to lack of funding before the end of the second full rotation. Various changes in the experimental design complicated comparisons because of increased use of chemical fertilizers on the mixed and stockless sections in 1967. Two years later, financial difficulties and disagreements over experimental direction splintered the coalition of researchers and volunteers. In 1970, the Soil Association sold off the farm to avoid bankruptcy. The Haughley experiment was over.

Nevertheless, at the end of the 30-year experiment the soil organic matter in the organic section remained twice that in the stockless (conventional) section. A decade later, in 1980, a follow-up survey of earthworm abundance and soil properties confirmed a lasting influence of the different treatments. Earthworm abundance remained almost twice as high in the organic section than in the stockless section. And the organic soil remained visibly darker, retaining a "crumb" structure that promoted infiltration, whereas the stockless soil formed clods and puddled when it rained.

Balfour retired from the Soil Association in 1985. She suffered a stroke and died in 1989 at the age of 90, shortly after being honored by induction into the Order of the British Empire. Though her contributions to organic agriculture were recognized in her lifetime, the march of agrochemical agriculture overran the practices she advocated.

Today, advances in microbial ecology are vindicating Balfour's views. It has taken decades for evidence and understanding to catch up with her insights about the complex, symbiotic relationships between plants and soil life. It is becoming clear that farming practices influence plant health and the nutritional profile of foods through affecting soil life—just as Balfour suspected when she argued that healthy fertile soils grew healthy plants that fed the health of livestock and people.

Balfour's legacy lives on in the growing movement of people promoting the value of whole, fresh foods grown from healthy soils as the foundation for human health. Only now, perhaps, the time is ripe for her once-radical ideas to take deeper root.

In thinking about how to characterize and assess soil health, we might consider how the medical community handles the concept of health. Most simply, health can be defined as being free of illness. In 1946 the World Health Organization took it further, adopting a broader definition of health as a state of physical, mental, and social well-being and not merely the absence of disease. While health is hard to measure directly, we do have telling indicators of human health—like temperature, blood pressure, and blood sugar.

Despite its complexity, at a general level the same holds true for soil health. The first-order indicators to consider in assessing soil health are relatively simple things: soil organic matter and the abundance, activity, and community composition of soil life. And these are the very things that suffer under conventional agriculture.

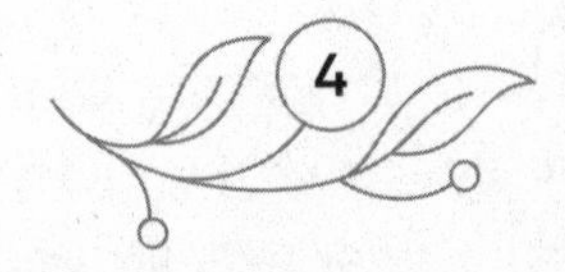

CONVENTIONAL DILEMMA

Gotta tend the earth if you want a rose.

—INDIGO GIRLS

Can we really farm to nurture soil life and build fertility yet still grow enough to feed us all? This is not a solely modern concern. Long before Balfour challenged conventional thinking, America's founding farmers worried about the state of their new country's soil. They saw fertile land as the cornerstone of democracy. Thomas Jefferson even cast the health of the soil as an intergenerational trust when he proclaimed, "While the farmer holds the title to the land, actually it belongs to all the people because civilization itself rests upon the soil."

In Jefferson's day, crop yields were down from what previous generations harvested from the eastern seaboard. Colonial plantation owners whose wealth was derived from the land and slave labor routinely complained about the degraded state of their fields. Jefferson wrote in his farm journal with growing concern about his soil, describing an interest in cover crops, crop rotations, and gully-filling to build and protect soils. He even bragged about inventing the finest plow. His new, curved-blade design turned the soil as it lifted, making it easier to work the land. The French Society of Agriculture endorsed Jefferson's immodest self-assessment and awarded him its gold medal.

Little did Jefferson suspect that his celebrated plow would plant a continental legacy of degraded soil. Only a decade after he died, blacksmith John Deere began marketing a new plow that improved upon Jefferson's design. Deere's steel plows could break through the dense roots that anchored the American prairie. Less than a century later the Dust Bowl's blossoming towers of wind-blown dirt revealed the price of plowing up the heart of America. Dramatic as this proved, the wake of the plow decimated more than the plains.

Were he alive today, Jefferson would likely lament the state of lands he knew well. The six or more inches of rich, black earth early European settlers described no longer caps the upland stretching from Virginia to the Carolinas. On farm after farm across the rolling hills of the Piedmont, reddish subsoil exposed at the surface reveals the true legacy of struggling farmers dependent on chemical fertilizers. Habitual plowing left once-productive fields bare and vulnerable to the erosive forces of wind and rain. Over time, topsoil slowly but surely bled off the land.

Underground Apocalypse

The plow is a double-edged sword for farmers and the soil. It's great for preparing the ground for planting, for weed control, and for producing a burst of fertility and growth as it temporarily hastens microbial breakdown of organic matter and the consequent release of nutrients. A commonly cited, though incorrect reason to plow is to get more of the rain that falls onto the ground to sink into the soil to better water crops and reduce erosive runoff. In practice, however, tillage makes it harder for water to seep into the ground because the pass of a plow breaks the natural conduits that carry water down into the soil. Plowing also pulverizes topsoil, producing a powder-like surface that crusts over when rained on. And when left bare, fields erode, carrying soil laden with fertilizers and pesticides off to cause problems elsewhere.

A plow also breaks up and disrupts the nooks and crannies in which soil life dwells. Under routine tillage the supply of organic matter declines, and soil life goes hungry. A review of more than 100 peer-reviewed papers in 1995 found that plowing profoundly disturbs soil food webs, especially the larger organisms. Loss of these organisms and their burrowing activity contributed to changes in soil structure, leading to more runoff and less water sinking into the ground.

Tillage also directly kills soil life. The most obvious and visible effect is on earthworms, which play a large role in nature's fertilizer system. The blade of a plow chops them up and destroys their burrows. This matters because worms are not just canaries in agriculture's coal mine. They help create natural drainage pathways in soil as well as subterranean habitat for themselves and many other kinds of soil life. Similar to how filter-feeding whales take in water, earthworms take in soil and sift through it to find food. They constantly mix together mineral soil and organic matter, enhancing fertility. As soil passes through earthworms, it becomes enriched with beneficial microorganisms and plant growth–promoting compounds produced in their gut. Earthworm activity also helps release nitrogen held in crop residue and soil organic matter.

In his final book, completed just months before his death, Charles Darwin cast worms as God's "ploughmen," marveling at how their tireless action recharged the soil and kept the English countryside productive through untold ages. They still do today. A 2014 global review found that earthworms boost crop yields by an average of 25 percent, enough to close the gap with conventional farming. This bit of welcome news means that worms can help feed the world without anyone having to eat one.

We're only now recognizing the devastating impacts of frequent tillage and intensive use of pesticides and chemical fertilizers on earthworms. A 2018 review of long-term farming trials—some run for over a century—found consistent losses of between half and all worm biomass, with an average loss of more than 80 percent. Apparently, modern farming methods have killed off the majority of worms that previously lived on farms.*

There really is no way around it. Plowing is a natural disaster for earthworms. The effects can be dramatic, with tillage depressing worm abundance even more than chemical fertilizers or pesticides. For example, a 20-year comparison of soil quality under conventional and no-till farming in New Zealand found conventionally farmed fields had virtually no worms, while no-till fields had almost as many worms as permanently unplowed pasture, which also had two to three times more microbes.

Tillage also is hard on mycorrhizal fungi. It slices through their extensive network of root-like hyphae, cutting off deliveries of everything flowing through them—minerals, water, and other compounds that benefit their plant partners. A 2016 review of 54 field studies found that less intensive tillage combined with cover cropping greatly increased mycorrhizal colonization of crop roots and the flow of nutrients from decaying plants and animals back into soil life and crops. Minimizing tillage is a way for farmers to cultivate more diverse fungal communities that boost plant health and productivity aboveground.

The effects on bacteria are more complicated. Bacterial communities can surge after tillage, as disturbance increases the surface area of soil and provides access to fresh organic matter to feast on. Occasional tillage proves a temporary disruption from which soil can usually recover. But regular tillage is more like a chronic disaster constantly

* It is worth noting, however, that in some areas non-native worms are an invasive species, like in Minnesota where they are damaging forests.

outpacing the accumulation of soil organic matter and impairing subterranean communities.

Tillage also affects the timing of nitrogen delivery to crops. A pulse of bacterial activity right after plowing makes more nitrogen immediately available. But it comes at the wrong time—well before the period of rapid growth when crops need it most or can take most of it up. The mismatch is a bit like trying to serve steak to a baby. Most of it ends up on the floor. So a lot of the nitrogen liberated from organic matter in tilled soil simply runs off and is lost from the fields. This problem is even more pronounced for synthetic nitrogen fertilizers as their high solubility leads to rapid leaching from the soil in early-season storms well before most crops can take them up.

So what's to be done about the devastating effects of the plow? The good news is that there are ways to farm that get seeds into the ground without ripping it up and turning it over. No-till planters slice a narrow incision into the ground through the mulch created from dead plant matter left from prior harvests and then drop seeds down into the underlying soil. Farmers call such layers of organic mulch "residue," though that term belies its true value. The combination of organic mulch and minimal soil disturbance provides ideal habitat for the types of soil life that improve agricultural soils and benefit crop health.

Farming systems that create and maintain high levels of soil organic matter work like a savings account, storing nutrients from one growing season to the next for the use of subsequent crops. This is what Eve Balfour realized early on. And combining no-till farming with cover crops will build up soil organic matter and deliver nitrogen more evenly across the growing season, kind of like delivering nutrients through an efficient drip irrigation system.

What difference can switching to no-till make? A 2018 comparison of conventionally tilled and no-till fields by a research team from Cornell University and the USDA showed that over a 12-year period no-till farming resulted in higher levels of soil organic matter, greater microbial biomass, greater levels of plant-available zinc in the soil, and less runoff, keeping more water in the soil where it helps crops grow. Introducing a cover crop further increased each of these positive effects, as well as the amount of plant-available iron in the soil. In this way, farming practices that increase soil organic matter increase the quantity of micronutrients crops take up.

Such results are not unusual. Reviews of the effects of no-till farming on soil organic matter report increases in the topsoil and mixed results for the full soil profile. While most such comparisons treat no-till as a stand-alone practice, reaping the full potential for increasing soil organic matter appears to depend on integration with other practices, particularly cover cropping.

Add compost, and the results are even more positive, as a University of California–Davis study found in comparing the effects of fertilizing with synthetic nitrogen, winter cover crops, and composted poultry manure on tilled corn-tomato and wheat-fallow crop rotations. Over the almost two-decade study, soil organic matter in conventionally fertilized fields did not increase. Adopting cover crops increased carbon in the uppermost foot of soil but reduced it deeper in the six-foot-deep soil profile. But overall soil carbon levels increased by about two-thirds of a percent a year using both cover crops *and* composted manure. The authors attributed this striking difference to compost feeding (and perhaps seeding) life in the soil.

Crop rotation—or the lack of it—also affects the diversity and composition of mycorrhizal fungi. Experiments at the University of Kentucky in the 1990s showed that crop rotations dramatically increased fungal diversity in fields with rotated crops compared to fields continuously planted with the same crop. The soil microbial community also shifted rapidly, but the effect varied depending on which crops were grown in what order. Sometimes mutualistic interactions flourished, and other times pathogenic ones increased.

When the same crop is planted over and over again, it acts as an invitation for pests and pathogens, whose next generations start life in a well-stocked grocery store catering to their particular tastes. But growing a greater variety of crops invites a diversity of microbial life, the vast majority of which are neither pests nor pathogens. Mixing up a sequence of different crops keeps pests off balance. That translates into less need for applying pesticides on food destined for our tables.

In the late 1990s well-replicated ecological field experiments in grasslands confirmed Charles Darwin's proposal that more diverse plant communities are more stable and productive. It is also well known that soil fertility and productivity influence plant diversity, with more productive sites supporting greater diversity. That this works both ways—diversity supports productivity and vice versa—suggests a positive feedback relevant to agricultural settings.

The most dramatic increases in soil organic matter and soil health occur with combining low- or no-till practices, cover crops, and diverse crop rotations. For example, a long-term study in southern Brazil documented that 25 years of conventional tillage decreased soil organic matter to less than a fifth of the amount in native soils. But soil organic matter recovered almost fully just two decades after switching to no-till, cover crops, and diverse rotations. We've seen firsthand similar results in Ghana, Ohio, Saskatchewan, and the Dakotas, where regenerative farms that adopted practices based on a combination of minimal disturbance, cover crops, and diverse rotations restored soil organic matter to levels close to or exceeding those in local native soils.

The full system works far better than individual practices to rebuild soil fertility. Reducing soil disturbance *and* providing food sources cultivates and protects beneficial soil life. Collectively, no-till systems with cover crops and diverse crop rotations mimic nested sets of ecological relationships between microbes, plants, and animals that allow each to thrive as part of a larger ecosystem. The classic example is one in which herbivores would starve without plants, carnivores would perish with no prey, and scavengers would have little to eat without the other two. Increasing soil organic matter sets the stage for more diverse and consistent food sources that lead to more diverse—and beneficial—communities of soil-dwelling organisms. But what happens if we feed the soil too much of just a few things?

Junk Food for Soil

The merits and side effects of nitrogen fertilizers have long been debated, often with great passion. Proponents of now-conventional methods often point to higher crop yields and jump straight to the without-chemical-fertilizers-we'd-all-starve argument. This point, however, is nowhere near as clear as usually asserted. Many of the studies that report lower production from organic farming systems reflect or include data from the transition period and the legacy of degraded, organic matter–poor soils.

At the other end of the debate about using nitrogen fertilizers is the argument that although they fuel plant growth, their damaging effects on soil life are too great to prove sustainable. These diametrically opposed views emerged from the inherent tension in agriculture between the volume harvested over the short run and the nutritional quality of a crop over the long run.

A go-to talking point of conventional interests centers on why it would matter whether nitrogen comes from a synthetic or an organic source. But that's the wrong question—for of course it doesn't matter. Nitrogen is nitrogen. Yet it matters—a lot—if nitrogen is combined with other elements that make it soluble enough to deliver at high rates. A 1974 review of the effects of two nitrogen fertilizers on plant diseases published in *Annual Reviews of Phytopathology* reported that the most common forms of nitrogen applied as fertilizers increased the occurrence and severity of different crop diseases. In addition, around half of the nitrogen applied as synthetic fertilizers runs off fields before crops can take it up, and overfertilizing undermines crop defenses, opening opportunities for disease and pathogens, with loss of mineral nutrients as a side effect.

Long ago the botanical world solved its nitrogen problem in a different way. Many plants partner with nitrogen-fixing bacteria. And microbial decomposition of organic matter also provides nitrogen during the growing season. As temperatures rise in spring and summer, microbial communities that decompose organic matter increase their activity, making more nitrogen available in the soil for plants to take up. This arrangement works quite well as it coincides with peak crop growth.

A key problem with long-term reliance on synthetic nitrogen fertilizers is that they stimulate microbial activity, speeding up the pace at which organic matter breaks down and releasing nutrients at times crops can't use them. Studies using fertilizer tagged with a tracer report that corn, wheat, and rice all take up more nitrogen from the soil—from soil organic matter—than directly from the fertilizer. So liberal applications of synthetic nitrogen fertilizer feed microbes that increase the pace of organic matter decay, thereby reducing levels of soil organic matter on conventional farms.

Of course, using soluble nitrogen fertilizers makes sense insomuch as you want plants to take up fertilizers. And a logical way to figure out how much nitrogen to add to a field in order to grow a particular crop is to measure the amount of readily soluble nitrogen in the soil and then compare it to the desired agronomic level. Most of the time this means advising farmers to add lots of nitrogen, so that's what they do.

But what farmers learn about when they get conventional soil test results is only what's already available to the plants rather than how much nitrogen-fixing bacteria might provide or decomposers might release from organic matter. So in the end the typical soil test protocol leads to routine overapplication of nitrogen.

We had this confirmed for us when we were speaking at a farming conference in Virginia, near Thomas Jefferson's old neighborhood.

Another speaker, a USDA researcher from the southeast, reported that the optimal amount of nitrogen fertilizer to apply to corn depended on how much biological activity and organic matter the soil held. His graphs, however, showed that when calculating recommended amounts of nitrogen fertilizer, agricultural extension agents in the region assumed soils held no organic matter. The presentation further revealed that the optimal nitrogen fertilizer application rate was zero for farms with the highest levels of soil organic matter and biological activity. In other words, applying synthetic nitrogen fertilizer proved a waste of money for healthy soil. It was solid evidence that rebuilding soil organic matter can greatly reduce the need for nitrogen fertilizer.

Soil acidification is another troubling long-term effect of relying on chemical fertilizers. A 2015 global review concluded that the soil became more acidic (low pH) in proportion to the amount of nitrogen fertilizer added to the soil. The resulting acidification can deplete mineral nutrients, like calcium, magnesium, and potassium, and increase the solubility and availability to plants of elements that produce toxic effects at high concentrations (like aluminum). In extreme cases, under continued use of synthetic nitrogen the soil can become so acidic that it retards root growth, harms soil life, disrupts nutrient uptake, and eventually poisons crops. Though a matter of degree, this is another factor potentially involved in nutrient declines in food. It is also interesting to note that acidic soils harm the nitrogen-fixing bacteria that form symbiotic relationships with legumes, which then increases the need for nitrogen fertilizers.

Too much readily soluble nitrogen also changes soil communities, with some studies reporting increases in total microbial biomass and others finding decreases. Yet more soil life is not necessarily better. What organisms are there and what they do makes all the difference. Plants depend on certain partners to acquire particular mineral micronutrients at the right time in a crop's growth cycle. And this charts a direct connection between farming practices and plant health.

It is well known that increasing the amount of nitrogen in soil through synthetic fertilizers reduces the abundance and diversity of mycorrhizal fungi. In particular, field-scale and experimental studies show that high nitrogen fertilizer use selects for less mutualistic fungi. So the more nitrogen we apply chemically, the less benefit our crops derive from fungal alliances.

In contrast, nutrients in the form of cover crop exudates and slowly decaying plant matter, as well as cattle manure, increase fungal abundance and diversity in soils. Mycorrhizal fungi can take up immobile elements like zinc and phosphorus and deliver them to crops. And other types of fungi and bacteria in the soil increase the availability of iron for plants to take up.

What is little appreciated in debates about conventional farming practices is that regular tillage and chemical fertilizer use disrupts the symbioses between plants and soil microbiota related to nutrient acquisition. With nutrients available for free from chemical fertilizers, plants can skip getting nitrogen, phosphorus, and potassium from soil-dwelling symbionts. But as with most free lunches, they aren't really free. Crops also lose out on other deliveries—from minerals to beneficial microbial metabolites and signaling compounds that stimulate plant defenses. There are other repercussions too. The flavor of crops can suffer, which, as we'll see, is more intimately linked to our health than you might suspect.

All in all, a shuttered root microbiome undercuts the botanical world's grand strategy for health—a robust, impenetrable defense. A key part of a plant's health plan relies on phytochemicals, many of which are produced when plants interact with soil life. And so while crops can grow robustly if routinely supplied with synthetic fertilizers, losing microbial partners disarms their defensive systems. This leaves them like a deer with a broken leg trying to flee from a pack of predators.

The recent discovery that root exudates feed microbes whose bodies become a primary source of soil carbon highlights the foundational nature of the reciprocal connections between plants and beneficial soil organisms. A 2017 analysis of 56 studies from around the world reported direct comparisons of paired conventional and organically farmed fields with the same soil type for an average of 16 years. Organically farmed soils held almost 50 percent more carbon and nitrogen in microbial biomass, with much greater microbial diversity and activity than conventionally farmed soils.

That the tiny bodies of exudate-eating microbes are major contributors to soil organic matter sets up one of the least recognized but most damaging effects of fertilizers. Crops raised on soluble synthetic nutrients cut back on exudate production. In effect, plants become botanical couch potatoes, slackers that don't bother to participate in the numerous symbiotic interactions that would otherwise provision them with an array of molecules and compounds their green bodies need for robust health. To grasp the scale of this impact, it's helpful to know that 20 to 50 percent of the carbon a plant acquires through photosynthesis can become root exudates that support soil microbiota. So agricultural practices that influence exudate production affect both the provisioning of beneficial microbial metabolites and levels of soil organic matter.

The new view of the soil as a biological bazaar made its way into a 2016 review in *Advances in Agronomy* (not exactly a hotbed of radical thought). The authors concluded that the plant microbiome is now "widely considered to be crucial for maintaining plant health in natural and managed ecosystems by increasing access to water and nutrients and modulating resistance to pests and disease."* In other words, plants grow best—and thrive—not when microbes are absent but rather when particularly beneficial communities of soil microbes are present.

This reality points to a not-so-secret downside of conventional agriculture. The way most farmers coax high yields from degraded soil

---

* Reeve, J. R., et al. 2016. Organic farming, soil health, and food quality: Considering possible links. *Advances in Agronomy* 137, p. 331.

is with heavy tillage and nitrogen fertilizer use. A singular focus on high yields means we feed plants the nutrients that fuel growth, overlooking the nutrients and compounds they need to safeguard all that growth. When farmers and their crops become addicted to synthetic nitrogen, it's hard to quit pesticides too. Herein lies the catch-22 of modern agriculture.

Weedy Problem

The other mainstay of conventional farming, genetically modified crops, inflames passions on all sides. Critics see genetically modified (GM) crops as dangerous threats to our health and the environment. Proponents portray them as the technical path to food security. As passionate and engaging as they often prove, such debates can distract from serious questions about some of the other effects of modifying crops to tolerate herbicides.

Originally, advocates argued that GM crops would dramatically boost crop yields and cut the need for herbicides. Yet corn yields in Europe, where GM crops are banned, and in the United States, where GM crops are standard, both increased at the same pace over recent decades. Likewise, a 2016 National Research Council report found no evidence that yields increased any faster in GM corn, cotton, or soybeans than in non-GM varieties. In other words, rising yields were coming from old-school crop breeding rather than from introduction of new GM crops.

Much of the ongoing GM debate revolves around crops engineered for resistance to the herbicide glyphosate, as was done for corn, soy, and cotton in the 1990s. A key attraction was that farmers could spray glyphosate over an entire field and kill only the weeds. By 2011, almost 95 percent of the American soybean harvest carried the genetic modification that conferred resistance to glyphosate. It offered an easy button for weed control. After patent rights expired in 2000, the price of glyphosate dropped by almost half. Now, each year over a billion pounds of glyphosate are applied to the world's farmland. It is not an exaggeration to say we are dousing our fields with it.

What does this do to soil life—and us? Far more, it seems, than initially thought, advertised, or admitted.

Proponents of GM crops also claimed they would reduce pesticide use. Widespread adoption of crops genetically modified to resist glyphosate did indeed reduce reliance on some much nastier herbicides. But it took just five years after glyphosate came into common use for resistant weeds to begin appearing and spreading across American farms. A 2012 survey of farmers found that almost half reported glyphosate-resistant weeds on their farms. Now the growing problem of herbicide-resistant weeds is spurring new efforts to market even more toxic multiple-herbicide cocktails. That's not progress. It's digging the hole you're in deeper.

So far it appears there is little direct evidence that GM crops themselves pose a serious health risk to the people or livestock that eat them, at least, that is, according to several high-profile reviews. The same can no longer be said in the case of glyphosate.

Curiously, glyphosate wasn't intended to be an herbicide. Initially, it was patented as a metal chelator—something that reacts with and binds up mineral elements. The idea was to use it to clean out metallic crud that had built up inside corroding pipes.* This may work great for cleaning crusty pipes, but a metal chelator added to soil binds to and ties up mineral elements like copper, iron, magnesium, and zinc, which makes them unavailable to plants and soil microbiota. In particular, glyphosate takes manganese out of biological circulation, thereby inhibiting an essential metabolic pathway for plants and many bacte-

* The original patent (U.S. Patent 3,160,632) for glyphosate was issued to Stauffer Chemicals in 1964. The patent for glyphosate as an herbicide (U.S. Patent 3,455,675) was issued to Monsanto Chemicals in 1969.

ria. Depriving plants of mineral elements central to their health and defenses is not the best move for a farmer—or those who eat their crops.

Arguments that glyphosate is perfectly safe rest on studies that found little to no direct toxicity to mammalian cells. But consider that Monsanto's second patent on glyphosate was for use as an antibiotic (U.S. Patent 7,771,736). With a flood of research and evidence showing how critical microbiomes are to their plant or human host and how antibiotics alter microbiomes, troubling implications come to mind. Imagine, for example, what this could mean for our health given the role of human gut microbiota in making compounds that modulate our metabolism, mood, and many other aspects of health.

While you may find the potential effects of glyphosate concerning, the Environmental Protection Agency appears undisturbed. In 2013 it raised allowable levels of the herbicide enough to ensure the food supply remained safely in compliance. Since then, new uses have emerged for glyphosate. It acts as a dessicant when sprayed on wheat right before harvest, making it much easier to separate grain from chaff. But doing so makes it virtually impossible to keep grain free of glyphosate.

Among the other concerns stemming from glyphosate use is that it can affect the nutritional quality of crops due to the mechanism through which it kills plants. The herbicide disrupts a key biochemical pathway.* Plants succumb not because of a toxic effect but because they can no longer make several amino acids that function as key building blocks for proteins and other compounds they need for growth, health, and defense.

In addition, glyphosate severely reduces the calcium content in the mitochondria of soybean root and leaf cells. While proponents tend to argue that glyphosate does not affect mammalian cells, studies in rats, nematodes, and zebrafish all found that exposure

* The disrupted pathway is the multistep shikimate pathway that plants, fungi, and bacteria use to make the amino acids phenylalanine, tryptophan, and tyrosine.

compromised or damaged mitochondria. Given the bacterial origin of human mitochondria—and the known antibiotic properties of glyphosate—it would seem advisable to minimize our exposure to chemicals that can damage the organelles that power every cell in our bodies. Such research undercuts the assumption that glyphosate can't harm us.

And glyphosate definitely affects earthworms. In an Austrian greenhouse experiment, earthworm activity virtually ceased several weeks after glyphosate-based herbicide applications. Within three months earthworm reproduction was reduced by more than half. Other laboratory experiments show that worms flee and avoid soil contaminated at concentrations recommended for commercial application of glyphosate-based herbicides and that such exposure may compromise earthworm survival even when concentrations result in low direct toxicity. Such effects are not limited to glyphosate. A five-year study in France showed that curtailing use of herbicides, insecticides, and fungicides could quadruple earthworm abundance.

Still, the effects of glyphosate on soil life are neither uniform nor simple. It is well established that repeated applications cause some soil microbes to die off while others can metabolize (eat) the herbicide. This explains why some studies report that glyphosate applications actually increase total microbial biomass. The right question, of course, is more nuanced: are these altered soil communities enhancing, harming, or having no effect on crop and human health?

Of particular concern is the effect of glyphosate on the beneficial microbial communities that live in and adjacent to a plant's roots. The herbicide shifts the community composition in ways that are generally detrimental to crop health. Plants themselves deliver glyphosate to their microbial helpers, as was found in soybeans that exuded the herbicide from their roots for several weeks after being sprayed. Delivering a broad-spectrum antibiotic to the rhizosphere—the home for microbial communities that provide nutrients to plants and keep pathogens at bay—is not exactly a recipe for improving soil health, crop health, or the nutrient density of food.

It's worth recalling here that the vast majority of land plants depend on mycorrhizal fungi. A 2014 review of the influence of glyphosate and other herbicides on soil biota concluded that herbicides in general harm beneficial soil fungi. And numerous studies have shown that glyphosate in particular decreases mycorrhizal colonization. Glyphosate has even been shown to affect mycorrhizal colonization of coffee plants enough to adversely affect their growth. Furthermore, some common soil fungi are quite vulnerable to glyphosate-based herbicides at application rates well below those typically used in conventional agriculture.

Laboratory experiments also show that glyphosate applications not only reduced populations of crop-beneficial bacteria but could completely suppress them, including those that solubilize and deliver phosphorus to their plant host. In one study, microbial deliveries of phosphate decreased with higher glyphosate application rates, dropping up to 84 percent. This suggests that greater herbicide use increases the need for fertilizer. The combined effects of reducing beneficial bacteria and promoting fungal pathogens map right onto Howard's and Balfour's suspicions about the downside of industrialized agrochemical agriculture.

Another Way

So is there a way to avoid the impacts on soil life that come from conventional farming? There is. The combination of no-till planting, cover crops, and diverse crop rotations can turn agriculture into a soil-building enterprise. Reintegrating livestock grazing into cropping systems can also help—if done right. Altogether these practices can transform today's high tillage, chemical-intensive monocultures into the regenerative farming of tomorrow. The good news is that research shows that this combination of practices can deliver positive results for farmers—and the land. We don't face a stark choice between agrochemical-intensive farming and mass starvation so long as we ditch the plow, keep the ground covered with living plants, and grow a diversity of crops.

Innovative farmers around the world—conventional and organic alike—are using regenerative practices to rebuild soil health and thereby boost soil fertility, productivity, and drought resilience. Cultivating beneficial microbial life is emerging as the new conventional wisdom. Farmers who adopt soil-building practices generally find that as their soil improves, they can spend less on fertilizers and pesticides, slashing some of their biggest costs without sacrificing high yields. And spending less on chemical inputs can mean the difference between healthy profits and financial insecurity.

A 2012 Iowa State University study compared the productivity and profitability of conventional corn-soybean rotations with regenerative practices that combined more diverse rotations, lower synthetic nitrogen and herbicide inputs, and periodic applications of cattle manure. The profitability and yields under regenerative practices were comparable to or greater than those of the conventional system, and weeds were effectively suppressed in both systems. Notably, the more diverse systems used less than a fifth of the synthetic nitrogen and herbicide, with a corresponding drop in nitrogen and phosphorus in runoff.

Experimental studies and recent meta-studies found that no-till methods, cover crops, and more diverse crop rotations increase microbial abundance, diversity, and activity. The combination makes for more resilient farming systems, reducing weed, insect, and pathogen pressures without reducing crop yields. But if you leave one or two of these three practices out of the mix, it doesn't work anywhere near as well. Another review of no-till practices found that they tended to produce lower crop yields—unless combined with cover cropping and crop rotations.

A particularly illuminating fourteen-year French study tracked the response of soil life to both conventional and organic farming, as well as a system known as conservation agriculture that combined no-till methods and cover crops and applied pesticides only after an economic damage threshold was exceeded. The conventional system used pesticides and mineral fertilizers, whereas the organic system used green manures (legume cover crops). Both the conventional and organic fields were plowed three out of every four years. Over the course of the study the abundance and biomass of soil life increased under both organic farming and conservation agriculture—visible soil life (earthworms and arthropods) increased 2- to 25-fold, and microorganisms (bacteria and fungi) increased by 30 to 70 percent. While conservation agriculture practices increased soil life across the board, organic practices mainly increased bacteria and earthworm populations. The study also found that combining no-till methods and cover crops is better for soil biota than just incorporating legumes in crop rotations or relying on pesticides and mineral fertilizers.

Today we have a much better understanding of the effects of practices like tillage on soil health. But we are far from the first to ask whether conventional farming is good for the land—and us. This question was very much on the mind of the English physician who spearheaded the Medical Testament presented at the 1939 meeting in Cheshire.

THE FARMERS' DOCTOR

The health of soil, plant, animal, and man is one indivisible whole.

—EVE BALFOUR

Lionel James Picton was born in Bebington, a few miles south of Liverpool, England, in 1874 and would eventually go on to inspire Eve Balfour. After graduating with three degrees from Oxford he worked at several hospitals before settling into a quiet rural medical practice in northwest England. Noticing that his sickliest patients ate the worst diets, he became intensely interested in nutrition and health. And this, in turn, led him to explore how farming practices affected nutrition. Along the way, he became secretary of the Cheshire medical committee and the driving force behind its controversial Medical Testament. In 1946, at age 72, he published *Thoughts on Feeding*, a book reflecting his lifetime of experience as a country doctor observing profound changes in the dietary habits of the people he served.

Picton was particularly concerned about the effects of food processing and the mechanized wheat milling used to produce white flour. He became convinced that processing grains to make white bread robbed an unsuspecting public of nutrition—and health. He thought growing crops with chemical fertilizers compounded the problem.

Picton lamented that his medical training did not include more about the effects of agricultural practices on food. In 1938, after years of treating common illnesses he thought resulted from an inadequate diet, Picton read an article about Sir Albert Howard's composting system and ideas on the nutrition and health of crops. Preventing illness through the right choice of food involved more than just what people ate. How farmers worked their fields—how they treated their land—affected what was in their crops and how that food nourished people or not.

Picton saw Howard's "happy symbiosis" between plant roots and mycorrhizal fungi as providing crops with "health, disease resistance and, last but not least, flavor."* But sure as he was of how plants and those who ate them benefited from this relationship, he had no idea what the plants did in return for the fungi. Though the nature of the relationship remained cryptic, Picton believed scientific understanding would catch up eventually.

In the 1820s the roots of certain plants were found to contain microscopic threads of living nonplant tissue, but it took until the middle of the century to recognize these living curiosities as fungal hyphae and until the 1880s to uncover their partnership with plant roots. The more scientists looked, the more they found such symbiotic associations—in pines, oaks, beeches, orchids, and grasses, as well as in crops including tea, coffee, wheat, rice, peas, beans, clovers, hops, and grapes. More rule than exception, the mysterious partnering of root and fungus was how plants worked.

Like Howard and Balfour, Picton believed that soil fungi bridged a critical linkage between decaying organic matter and plant nutrition. Yet thinking that crop health depended on fungi and humus remained

* Picton, L. J. 1946. *Thoughts on Feeding*. Faber and Faber, London, p. 36.

heresy among agronomists. The standard view considered microbes bad and decaying organic matter putrid. What good could possibly come from germs?

But Howard's experience intrigued Picton. Compost had improved crop health and productivity on tea estates far too rapidly to be due simply to changes in soil chemistry. Just as interesting to him were Howard's observations of how crop health declined after the introduction of chemical fertilizers on tea estates in India. Time and again Howard saw fungal parasites promptly attack degraded tea fields when chemical fertilizers were introduced to boost yields. Sugar plantations experienced similar problems. Soils dressed with well-made compost supported healthy crops with abundant mycorrhizae, but chemically fertilized soils held hardly any. Here was evidence enough for Picton that connections existed between crop health and farming practices.

Convinced that Howard was right to think chemical fertilizers disrupted beneficial soil life, Picton suspected this also affected the quality of the food that grew in the soil. As evidence for the nutritional superiority of crops grown on manured fields over those treated with chemical fertilizers, Picton pointed to an experiment conducted by Moses Rowlands and Barbara Wilkinson of Knightsbridge Laboratories. Rowlands had noticed that pigs fared better on home-grown barley and wheat meal than on chemically fertilized crops. Puzzled, the pair wondered whether the difference in the growth and health of the pigs was due to feed grown with chemical fertilizers being nutritionally deficient in some way.

So they planted grass and clover in a field that previously had been used to grow cabbages, potatoes, and wheat. One half of the field was fertilized with "artificial manure" (chemical fertilizers), and the other half was fertilized with manure from pigs on the barley and wheat meal diet. The grass and clover seeds were then harvested and fed to two groups of rats, one receiving the manured seeds and the other the "artificially manured" seeds.

For a week and a half all the rats grew at about the same pace. Then the rats eating the chemically fertilized seeds became sickly and stopped growing. Those eating the seeds grown with manure continued to grow as before. Something essential was missing in the diet of the rats eating the chemically fertilized seeds. Rowlands and Wilkinson thought it might be the recently discovered vitamin B (now known to consist of a complex of multiple vitamins).

Their second experiment involved feeding two groups of rats an identical, vitamin B–deficient diet for 18 days and then adding seeds grown on manured soils to the diet of one group and chemically fertilized seeds to the diet of the other group. After a week on the vitamin B–deficient diet both groups began to lose weight. The group that then received manured seeds recovered rapidly and began to grow normally. The group that received the artificially manured seeds kept declining. By the twenty-first day they were wasting away and covered with unsightly bald patches. With death looking imminent for these unfortunate rodents, Rowlands and Wilkinson switched their diet to manured seeds. Within a day, the rats revived and began growing normally again. Apparently, the pig manure passed vitamin B to the soil, and the crops absorbed it and passed it on to the rats. How this causal chain worked they had no idea. But the nutritional value of the manured and chemically fertilized crops differed. They concluded with an understated comment that still rings true today: "If these results are correct, they have a considerable bearing upon agriculture and nutrition."*

Around the same time, agronomists in Ohio studied the effects of chemical fertilizers on the nutritional value of wheat. Finding little difference in vitamin B levels of wheat grown with different types of chemical fertilizers, they did not bother to test whether crops fertil-

* Rowlands, M. J., and B. Wilkinson. 1930. The vitamin B content of grass seeds in relationship to manures. *Biochemical Journal* 24, p. 204.

ized with manure held comparable levels of vitamin B. Why test yesterday's practices? American agriculture was marching forward, not looking backward.

Picton wasn't so sure. He described the experience of a farmer whose chemically fertilized farm had degraded soil and diseased cattle. After seeing a sickly mare recover rapidly upon being switched to manured pasture, the farmer began making his own compost and growing plants without chemical fertilizers to feed his livestock. Soon the health of his livestock improved dramatically, as if they had an inherent drive to heal. Within a few years they rebounded back to disease-free robust health without vaccines or medicines.

To Picton these were clear lessons. Proper feeding was the neglected key to robust health. And the foundation for proper feeding? Healthy soil. Pathogens were always around, but animals better resisted diseases when grazing pastures with healthier soil.

Picton also saw that food processing affected people's health. In particular, he pointed to the results of a 1927 Swiss military manpower review that found the occurrence of goiter, a swelling of the neck from enlargement of the thyroid gland, to be far more common in German-speaking cantons than in French- or Italian-speaking cantons. The root cause lay in different cooking customs. German speakers discarded the water their vegetables were boiled in. French and Italian speakers used it to make soup. The vegetable water provided enough iodine boiled out of the vegetables to protect the soup eaters. Tossing the vegetable water robbed the German speakers of an essential protective nutrient.

Picton was also particularly concerned about what happened to food before it reached the kitchen. Few were as aware as he about the way the industrial revolution transformed the English diet in the late nineteenth century. From Norman times to the 1870s, the wheat that made up England's daily bread was ground on stone mills. Hearty whole-wheat loaves fed the working poor. It took a lot of labor to sieve and sort out flecks of dark bran to make fashionably clean white flour. White bread was expensive, a luxury for the rich.

In the great rush of industrialization, steel roller mills replaced stone mills and mass-produced cheap, white flour, efficiently separating the dark fiber and vitamin- and mineral-rich bran and germ from the lighter, starchy endosperm. The main reason for sieving the color out of bread was pure fancy—making what had once been a frivolous luxury available for the masses. Traditional brown-colored, nutty-tasting, whole-grain bread was trampled in a stampede of progress.

Picton remained perplexed by the attraction. He found the tasty, filling loaves made from whole-meal flour far superior to the fluffy, insubstantial white bread that filled store shelves. The key advantage of the modern loaf lay elsewhere, in other, more commercial considerations.

Millers loved the new white flour. Removing the germ removed the pesky fats that led to a short shelf life. Shippers concurred, as flour shipments could now travel farther or sit around longer before spoiling. And while eating dark bread carried social stigma, consuming white bread conveyed social standing long reserved for the aristocracy. So now that anyone could eat it, everyone wanted to.

Not that they should, for the difference was striking. In 1930, Picton confirmed that processed flour stripped vitamins and minerals from the English diet through chemical tests conducted on loaves of whole-grain and white bread. Whole-wheat bread had three times the iron, twice the phosphorus, and seven times more vitamin B. White bread had no vitamin A, whereas whole-wheat bread held even more vitamin A than vitamin B. The pattern was clear. Stone-milled, whole-grain flour contained far more vitamins and minerals. Processed, steel-milled white flour was almost pure starch. Picton considered this basic change in the nutrient density of a main staple of the English diet a public health disaster hiding in plain sight at every bakery throughout the kingdom.

Experiments backed up his opinion. Sir Frederick Keeble gave three groups of caged rats access to both whole-meal and white flour. Each group ate three to six times more whole-meal flour than white flour. The rats not only preferred whole-meal, they grew in direct proportion to how much of it they ate, gaining almost twice the weight per pound than when consuming white flour. Whole-meal flour provided greater nutrition. Picton concluded that the switch to processed white flour unwittingly sabotaged English health.

How did this work? Along with A and B vitamins, vitamin E is concentrated in the wheat bran and germ rather than the endosperm (kernel) that becomes white flour. So while removing the wheat germ and the problematic fats it contained helped keep flour from spoiling and increased transportability, consumers paid a price in nutritionally diminished food for the commercial advantages flowing to millers, shippers, and merchants. After the development of steel roller mills, up to 30 percent of the weight of whole grains was removed during processing to make white flour.

Picton connected the accompanying loss of nutrients to compromised public health, pointing to experience in the First World War that showed whole-meal grains helped prevent debilitating diseases. During the Ottoman army's 1916 siege of Kut Al Amara 100 miles south of Baghdad, the British and Indian defenders suffered from different maladies due to differences in their rations. British troops were issued white-flour biscuits and fresh horsemeat. Indian troops were issued whole-wheat and whole-barley flour and refused to eat horsemeat for religious reasons. Only the British troops suffered from beriberi, a serious disease that leads to circulatory, muscle, and nervous system disorders due to vitamin B_1 (thiamine) deficiency. To Picton, an overarching lesson stood out. Processing grains to make white flour robbed food of nutrition.

He thought that this had also happened with rice. Just as for white flour, milling of rice removes the outer layers of the seed coat, including the bran and the inner germ, leaving the starchy endosperm. The development of modern technology to mill and polish rice stripped dietary fiber, healthy fats, vitamins, and minerals from a key part of the human diet. In Picton's time, it was well known that beriberi spread across Asia when shiny, pure-looking white rice replaced brown rice as a regional staple.

Yet white rice is not a modern invention. Picton described accounts of manually milled white rice in ancient China. Because milling costs money, white rice was more expensive and came to be perceived as a superior good. When industrial-scale milling machines delivered this luxury item to the masses, they eagerly adopted it as brown rice became passé, seen as dirty and inferior.

Another reason for the rapid adoption of white rice lay in the grain's inherent properties. Brown rice, like most whole grains, contain fats that go rancid unless properly stored. Nineteenth-century traders demanded polished white rice for shipments because it lasted longer before going bad—it shipped better.

McCarrison conducted experiments that confirmed Picton's fears. Pigeons fed a white rice–only diet developed beriberi-like symptoms and rapidly progressed toward death. Yet they fully recovered within hours if simply fed the bran and germ that had been removed from the white rice they'd been eating. McCarrison's pigeon experiments demonstrated the nutritional inferiority of white rice. He considered it no accident that areas of India where people consumed brown rice reported ten times fewer beriberi cases than those where people ate polished white rice.

Decades before, in the 1870s, the Japanese navy learned a parallel lesson when beriberi swept through more than a third of its sailors. Takaki Kanehiro, the fleet's surgeon general, suspected a nutritional deficiency might be to blame. So he experimented with improving the diet for seamen, adding a little meat and vegetables to their mostly polished white rice ration. Soon the fleet sailed beriberi free. But the underlying nutritional lesson did not catch on. At the time, the medical profession considered beriberi an infectious disease.

The main problem with a white rice diet comes with relying on just white rice. Eat it with enough meat and vegetables and beriberi does not arise. So beriberi remained mostly a disease of the poor, who increasingly aspired to eat—you guessed it—white rice.

Picton's concerns about white rice remain pertinent today. A 2010 study by doctors at the Harvard School of Public Health found that higher intake of white rice was associated with a higher risk of type 2 diabetes and that consuming the same amount of brown rice lowered diabetes risk. The study reviewed decades of data from thousands of patients. Eating white rice five or more times a week increased the risk of diabetes by almost 20 percent compared to eating white rice less than once a month. Eating brown rice at least two times a week decreased the risk of diabetes by about 10 percent compared to eating it less than once a month. Overall, diabetes decreased with increasing consumption of whole grains. Similarly, a 2007 study of more than 60,000 middle-aged Chinese women found that greater white rice consumption increased diabetes risk.

In his day, Picton also noted how pellagra, a long-dreaded disease now known to arise from niacin (vitamin B_3) deficiency, commonly led to the three Ds—diarrhea, dermatitis, and dementia—in the southern United States, where many people ate a diet of cornmeal, molasses, and salt pork. He pointed to the example of South Carolina, where from 1911 to 1916 more people died from pellagra than from tuberculosis or malaria, two common and much-dreaded diseases at the time. Across the American South, between 1906 and 1940 more than 100,000 people died of pellagra. Noting that refining removed vitamins, Picton suspected that not refining corn, the backbone of the regional diet, would prevent pellagra deaths.

Across humanity's three staple grains, modern processing stripped critical nutrients from the bulk of our caloric intake. While the particulars differed for wheat, rice, and corn, the implications did not. Whole grains provided things that highly processed ones lacked. And these things mattered to health.

We now understand the roles various vitamins play in preventing diseases like beriberi, scurvy, and pellagra. Supplementation of processed foods with vitamins made them maladies of the past across the developed world. Today, U.S. law requires B vitamins and iron to be added back into white bread and rice. You can see this in the labeling of "enriched" flour and rice. Yet one can't help but appreciate Picton's conclusion that it would prove wiser to simply leave vitamins and minerals in grains in the first place. What else are we missing in our food today, and how much of a role did adopting modern farming practices play in it?

Following Clues

Picton did not stop at considering how raising crops influenced human health. He also thought about the quality (and flavor) of meat in the context of forage quality. Eating nutrient-rich organ meats like kidneys and liver was an excellent way for people to get vitamins as the animals had already acquired and concentrated them through their diet. Nutrition flowed from the soil to crops and livestock and on to people—but only if we grew and raised them in ways that ensured the transfers occurred.

While he could not offer a detailed explanation for how nutrients and other compounds moved from one source to another, Picton maintained that the onus was on the "chemical advocates" to demonstrate there were no deleterious effects on soil and crop quality from synthetic fertilizers. After all, Eve Balfour reputedly could determine with certainty from a single bite whether a tomato had been grown with compost or chemical fertilizers.

In 1948, two years after publishing *Thoughts on Feeding*, Picton died at the age of 74. His obituary in the *British Medical Journal* called the Cheshire Medical Testament a "monument to his diligence and enthusiasm." But his insightful book did not merit notice, as the journal posthumously labeled him "an able writer, fresh and often provocative, though he never published a book." That oversight was kinder than the reception his book received in the *Quarterly Review of Biology*, which reviewed the just-published American edition a year after Picton died.

The reviewer, Johns Hopkins University geography professor Robert Pendleton, argued that in challenging the use of chemical fertilizers Picton was overlooking that scientists had established that healthy crops also needed mineral micronutrients—and that these too could be added along with major element fertilizers. The problem wasn't using chemical fertilizers but not using enough of them. Farmers just needed to add the right mix of chemicals to dial in optimal nutrition. Pendleton glibly ignored Picton's argument that soil biology was intimately linked to plants taking up nutrients, and he went on to argue that manure supplies could not possibly keep up with growing demand for fertilizers. His review concluded with an attack on those who would argue that farming methods affected food quality. People needed enough to eat and food was food, however one grew it. In the modern world, quantity trumped quality and that was that.

Surprisingly, Picton received a more insightful posthumous review in the *Agronomy Journal*, a mainstay of conventional farming interests. After noting that many of the journal's readers would disagree with the contention that chemical fertilizers were bad for the soil, the reviewer also noted that anyone who gave the subject serious thought would agree that the way farmers treated their soil was fundamental to growing nutritious crops. And the mycorrhizal connection that Picton pointed to might help explain why certain plants grew better in certain soils. The effects of food processing and cooking methods were known to affect human nutrition. So might not farming practices and the state of the soil do so as well?

The English experience during the Second World War inadvertently made Picton's case, demonstrating the beneficial influence of a better diet on public health. Although geography protected the British Isles from invasion, it rendered them vulnerable to blockade. Before the war Britain imported most of its food. Yet when the war halved imports, the British did not starve. Unlike most of continental Europe, public health actually improved during the war years.

The plan that produced this remarkable result sprang from the mind of Jack Drummond, a debonair scientist born in 1891 who was appointed scientific adviser to the Ministry of Food in February 1940. He was the right person for the job.

Two decades before, Drummond accepted a post as a reader at University College, London, and made a name for himself proposing that the vital substances known as "vitamines" should be modified to "vitamins" as some of the newly discovered substances were not actually amines.* He also introduced the now-familiar alphabetical labels of A, B, C, and so on to distinguish among vitamins known at the time. Through the 1920s Drummond made fundamental contributions to understanding their structure and importance to health. After being appointed the university's first professor of biochemistry, he increasingly turned his attention to nutritional questions that paralleled Picton's concerns about the effects of milling on the nutritional value of bread.

In 1939 Drummond published *The Englishman's Food* with Anne Wilbraham, his research assistant (and soon to become second wife). In exploring the history of the English diet, the pair emphasized the role of diet in combating disease. Throughout history the poor had too little to eat while the rich ate too much. Of particular concern was the declining quality of food available to the urban poor. Vitamin deficiency was widespread, with little to no vitamin A or D in the white

* Amines are organic compounds derived from ammonia (NH_3) with one or more of the hydrogen atoms replaced by an organic functional group.

bread the working class subsisted on. Dietary intake of vitamin B_1 was just a third of that in preindustrial times. Drummond and Wilbraham noted that at the start of the twentieth century malnutrition was more widespread in England than at any time since medieval days. The message? What one ate and how it was prepared were central to health. A review in *Nature* called the book "stimulating beyond words" before concluding that "we ought to know a lot more about the inner make-up of our foodstuffs."

During the war Drummond faced the challenge of how to adequately feed a nation cut off from much of its prior food supply. Yet he set his sights higher, designing a rationing system to combat dietary ignorance as he sought to improve and not just maintain public health.

His system directed more foods rich in protein and vitamins for the poor and less consumption of meat, fat, sugar, and eggs for the well off. With the ability to decide what foods should be imported and to incentivize maximal use of homegrown foods, Drummond reshaped the British diet, at least temporarily. He cut sugar imports to nineteenth-century levels and increased imports of cheese, fatty fish, and dried pulses. People were issued ration books that contained coupons redeemable for certain types of food and were encouraged to grow their own vegetables in public garden allotments in a "Dig for Victory" campaign. Drummond's plan emphasized supplying cod liver oil and vitamin tablets to nursing mothers and children. Most controversially, he stressed the need for more-nutritious bread, mandating that bakers revert to less-refined, closer-to-whole-wheat bread to make a "wartime loaf" that retained more vitamins and minerals.

Drummond's wartime diet plan delivered. Dentists noticed fewer cavities. Rates of anemia decreased among children. Infant and maternal mortality dropped to the lowest levels on record. So did stillbirths. Children grew taller, and diet-related diseases declined across the populace. Partway through the war, in 1944, Drummond was knighted for his service in nourishing Britain through a time of scarcity. In 1947, the American Public Health Association cited Drummond's wartime contribution as "one of the greatest demonstrations in public health administration that the world has ever seen." His obituary in *Nature* cited the remarkable achievement of improving public health during acute food shortages.

The effects of changing the English diet effectively proved Picton's argument that what people ate set the stage for health, good or bad. But what about Picton's other concern—the effects of modern farming practices on the nutritional value of food?

Too Much of a Good Thing

For some aspects of our diet it's not a shortage but rather too much that becomes a problem. Of those, nitrogen tops the list, especially in the developed world.

From 1945 to 1993 the use of nitrogen fertilizer in the United States increased 20-fold. Nitrogen in runoff and groundwater from agricultural lands is now the leading contaminant affecting public and private drinking water sources in the United States. A U.S. Geological Survey study of groundwater around the country found nitrate in more than 70 percent of samples, with almost a fifth of private wells exceeding maximum regulatory levels. Not surprisingly, shallow groundwater beneath agricultural land had the most. This problem is not limited to the United States. High levels of nitrates in drinking water are common in Europe, rural India, and the Gaza Strip, among other places.

Vegetables can be another source of dietary nitrate, as certain ones—particularly spinach, lettuce, Swiss chard, radishes, and cruciferous vegetables—accumulate it in their edible tissues when grown under high nitrogen fertilizer applications. But the level to which nitrates accumulate in plants depends on the form in which the nitrogen is applied. Highly soluble chemical fertilizers lead to higher soil nitrate levels compared to slow-release sources such as organic matter and compost. This is why produce grown with chemical fertilizers tends to have higher nitrate levels.

High levels of nitrogen in our food are associated with a range of ailments. Epidemiological studies have linked stomach cancer to dietary nitrate, as well as to exposure to nitrogen fertilizers. Why the connection? Bacteria in the mouth convert dietary nitrates into nitrites that react with amines in an acidic environment like the stomach to yield nitrosamines. One is NDMA (N-nitrosodimethylamine), a potent carcinogen that spawns tumors when fed to rats and monkeys. Nitrosamines have been found to cause cancer in more than three dozen species of animals, especially primates. Experiments on rats found that repeated dietary intake of even small amounts of nitrite consumed along with amines could prove carcinogenic and more dangerous than ingesting single large doses of either. Yet oral and stomach bacteria also convert dietary nitrates to nitric oxide, which at optimal levels is now understood to play beneficial roles in protecting against ulcers, stroke, and hypertension. While understanding the health effects of dietary nitrate consumption is complicated, it's fair to say that drinking water with excess nitrates is not healthy.

Studies of nitrates in drinking water supplies report consistently adverse human health effects. A study of cancer mortality rates in several hundred municipalities in a part of Spain where the concentration of nitrates in drinking water is the highest in Europe found people age 55 to 75 who consumed the most nitrate had twice the risk of stomach and prostate cancer—and a 40 percent higher risk of bladder cancer in men. Similarly, a more than decade-long epidemiological study of almost 22,000 Iowa women between the ages of 55 and 69 found nitrate intake associated with both bladder and ovarian cancer but not other types of cancer. Most of the nitrates one consumes pass rapidly to the bladder for excretion in urine. It makes sense that the places where nitrates are converted into known carcinogens are organs for which greater dietary nitrate consumption increases risk of cancer.

The authors of a 2018 review of epidemiological studies reported many other health risks for drinking water laced with nitrate, even at levels below regulatory limits. The review concluded that colorectal cancer and thyroid disease had the strongest epidemiological evidence for adverse health effects from elevated nitrates in drinking water. The review also reported that numerous studies found high nitrate levels in drinking water almost doubled the likelihood of birth defects involving the brain or spinal cord.

A Korean study looked specifically at the health risks of excessive nitrogen consumption due to NDMA and a diet rich in amines, organic compounds that result from the breakdown of proteins and from burning meat, among other sources. The study gave 40 volunteers a regular dose of nitrate and an amine-rich diet. Consuming whole strawberries, garlic juice, and kale juice, however, reduced NDMA formation by almost half to more than two-thirds. Vitamin C, polyphenols, and sulfur-containing compounds in garlic were particularly effective at reducing the production of carcinogenic nitrate-derived compounds. This helps explain why consuming fresh fruits and vegetables exerts a protective effect against some types of cancer: they help stop it before it starts.

It turns out that applying a lot of nitrogen fertilizer to fields also influences crop uptake of other minerals that plants need. Studies in the 1940s found that nitrogen and potassium fertilizers decreased calcium uptake by vegetables. Researchers in that era also found that nitrogen fertilizers reduced vitamin levels in grapefruit juice and pepper pods but increased vitamin C in a few other crops. While the mechanisms behind the influence of chemical fertilizers remained cryptic, the effects were real.

Turnip greens were popular to study at the time as they were commonly grown and a part of the region's diet. One study in Mississippi found that synthetic nitrogen fertilizer reduced their calcium content by more than a third in four out of five trials. A more comprehensive study in the same time period found similar results and declines in phosphorus as well. In addition, nitrogen fertilizer applications decreased the iron content of turnip greens by 20 percent on average in 26 of 29 locations across the American South.

While the size of the effect varied between locations, its direction was consistent. High levels of iron in turnip greens were associated with high soil organic matter—the only soil factor that significantly improved the iron content of turnip greens. In short, more soil organic matter translated into more iron in a regional vegetable staple.

A 1946 study in New Zealand investigated the effects of nitrogen fertilizers on dietary sources of several nutrients thought to have protective value for health. Researchers grew spinach under greenhouse conditions in the same soil supplying 4 different levels of both calcium (calcium acetate) and nitrogen (ammonium nitrate) for a total of 16 trials, each of which was replicated 10 times. As expected, yields increased with higher rates of nitrogen fertilization. But the vitamin C and phosphorus content of the spinach decreased with higher nitrogen applications. The researchers concluded that the spinach "became progressively inferior in nutritional value with each additional increment of nitrogen." The more they goosed yields with chemical fertilizers, the worse the nutritive value of food became.

Early studies pointed to contrasting effects of organic matter and chemical fertilizers on soil life in influencing the nutritional composition of crops, but understanding why the effects differed lay beyond the realm of conventional thinking. It took decades for science to unravel the mechanisms.

Yet evidence that synthetic fertilizers affect nutritional composition has persisted over subsequent decades. A 1993 global review found that nitrogen fertilizers significantly influence the vitamin content of plants, typically in unfavorable ways. After noting that the effects of nitrogen fertilizers received little attention in English-speaking countries after these early studies, the Swiss author reported that there was a rich body of work published in non-English-language journals that seemed to have gone "mostly unnoticed by English-speaking scientists."

This same review found substantial evidence that typical high rates of nitrogen fertilizer use greatly decreased the vitamin C content of many fruits and vegetables. They also reduced levels of vitamin E. Although the magnitude of the vitamin C loss varied for different crops, the decrease was quite significant, with reported reductions of up to 34 percent in cabbage and 50 percent in fruits. Conversely, lowering application rates of nitrogen fertilizers tended to increase the vitamin C content of crops.

In addition, the study author found that nitrogen fertilizers generally increased the protein and nitrate content of crops. This combination of reduced vitamin C content and increased nitrate content has other effects too. Vitamin C inhibits the digestive production of carcinogenic compounds from nitrate, and so lower levels of vitamin C would lessen such protective effects. The author urged that this relationship deserved further investigation as other studies suggested nitrate levels could increase to "potentially dangerous levels."* While the study also concluded that nitrogen fertilizers increased the vitamin B_1 (thiamine) and carotene content of crops, the author considered these increases nutritionally inconsequential for grains given that milling removed the bran and other vitamin-rich parts of the grain anyway.

For decades, the nutritional variability of crops was ascribed to different cultivars and underlying soil types rather than to how farmers treated their land. This thinking reinforced the view that effects of tillage or chemical fertilizers were not important. The headlong pursuit of higher yields made it hard to acknowledge, let alone to address, such concerns.

Yet as our understanding of connections between farming prac-

* Mozafar, A. 1993. Nitrogen fertilizers and the amount of vitamins in plants: A review. *Journal of Plant Nutrition* 16, pp. 2,480 and 2,487.

tices, soil health, and the nutritional quality of food grows, we've discovered that modern crops now pack less of some key things our bodies need. No doubt Picton, Balfour, and Howard would take an interest in what now-conventional farming practices do to the types and amounts of phytochemicals in our crops—and what these botanical gems do for our bodies when we consume them.

PLANT

BOTANICAL BODIES

Education is understanding relationships.

—GEORGE WASHINGTON CARVER

Is soil even necessary to grow healthy food? Not long ago no one would ever have thought to pose the question. Yet the issue became both relevant and controversial once soil-less hydroponic growers managed to convince the USDA that their methods complied with the National Organic Program rules and standards. Not surprisingly, some of the farmers who originally helped create the standards insisted otherwise in light of the clear language in the rules requiring soil health practices. In protest they banded together and formed a new, pointedly named breakaway organization—the Real Organic Project.

These folks could not imagine that any farmer, least of all an organic farmer, would want to grow tomatoes and lettuces in rows and rows of soil-free plastic containers designed to prop plants upright and bathe their roots in liquid nutrient solutions delivered through pipes, sprayers, and pumps. To them that pretty much undercut the whole point of establishing definitions and standards for organic farming in the first place. It just didn't seem necessary to actually spell out that you had to grow organic crops in soil.

How times change. Interest in going hydroponic exploded among industrialized organic vegetable operations once the USDA condoned the practice.

With no organic matter or soil life through which nutrients cycle, it's up to hydroponic growers to determine what's in the water their crops receive. So if minerals are low, they can add what's needed in soluble forms for plants to suck up. But first you'd have to find out what's missing, which takes time and money. Then there would be a decision to make: does one add the full spectrum of micronutrients that support plant (and human) health, or only the NPK that supercharges growth and yield? The historical behavior of conventional farmers suggests what one might expect.

A related question is whether phytochemical levels skew lower in hydroponic crops. There isn't much research on this question. But based on why plants make and use phytochemicals, it's reasonable to suspect that moving crops indoors and growing them without soil—and soil life—would alter their phytochemical profile.

Botanical Alchemy

Phytochemicals are a key reason that soil health matters to plant health—and ours as well. But don't go thinking that plants make phytochemicals because they're good for us. It's all about them. Every plant on the planet, from redwood trees in the wild to crops like wheat and watermelon, makes phytochemicals. Why? Land plants are stuck in place, and when they fire up their phytochemical factories, it stocks their green bodies with a pharmacy *and* an arsenal—time-tested botanical medicines and weapons essential to their health and defense.

Among the cues plants rely on to make phytochemicals are stressors like temperature, drought, and toxins. Biological interactions play a big role too, from the nibble of an herbivore to complex chemical signaling between a plant and the multitudes composing their root microbiome. Plants use phytochemicals as part of their sophisticated communication system, releasing them into the air and soil to warn their kin of nearby pests and pathogens.

Plants would be sitting ducks without producing phytochemicals and concentrating them in their green bodies when and where needed. And so crops that thrive despite the challenges and hardships of being rooted in the ground generally deliver higher levels of phytochemicals to our bodies. This ripples through to livestock too. The plants they eat set the phytochemical levels in the animal foods we eat. What our food ate shapes what our bodies get to work with.

Phytochemicals also lend colors to food. They tint the skin of an apple yellow, green, or red. They impart the vibrant red of a raspberry and the drab olive green of an artichoke. The phytochemical beta-carotene that makes the orange and yellow colors of squash works as a sunscreen, shielding plants from the damaging downside of intense sunny days. Beta-carotene serves the same purpose in us and is a common ingredient in sunscreen products.

Flavonoids are a type of phytochemical found in many plants and have a wide range of biological effects reflecting a long history of interactions between plants and their foes and allies over the course of evolution. Some have antifungal properties that keep harmful fungi at bay, and others help protect plants from insect and mammalian herbivory. Certain flavonoids orchestrate the most important and critical relationship of all—nitrogen fixation. Plants use exudates rich in flavonoids to draw the specialized bacteria that carry out this biochemical wizardry directly into their roots, or the rhizosphere. Once in place, the bacteria begin converting gaseous nitrogen into forms plants take up. When plants obtain nitrogen this way, it comes with some major advantages too. Farmers avoid steep fertilizer bills, and nitrates don't run out of people's taps.

Other types of flavonoids attract rhizosphere-dwelling microbes that solubilize iron, copper, and zinc, allowing plants to slurp up more minerals in the water they draw from the soil. Yet another type of flavonoid, benzoxazinoids, triggers complex defensive actions that repel a range of pathogens and insect pests in grasses, including corn, wheat, and rye. Still others, like tomatine, a glycoalkaloid, enhance fungal resistance in tomato and pea plants.

Seedlings are especially vulnerable to herbivore pests. Rapeseed youngsters produce glucosinolates that help repel slugs. Their brassica cousins, like broccoli, Brussels sprouts, and kale, deploy sulforaphane, another glucosinolate. This slick, James Bond-esque phytochemical acts like a bomb that insects themselves detonate. Biting into a leaf unleashes a chemical reaction that creates sulforaphane and sends insects scurrying off to chow down on something less noxious.

Yet phytochemicals can be inviting too. Thank phytochemicals for the spicy bite of fresh arugula and the sweet burn of chili peppers. Connections between flavor and many phytochemicals are simple and direct. This, perhaps, is why Eve Balfour could so readily distinguish between chemically fertilized and manure-fertilized tomatoes.

Today, our understanding of what is essential for health is changing as we learn about the beneficial roles phytochemicals play in plants and the parallels that carry over to people. A wide range of studies over recent decades shows that farming practices from crop breeding to agrochemical use influence the types and amounts of phytochemicals plants "decide" to make. In regard to nitrogen fertilizer application, research shows that it can reduce phytochemical production related to plant defense. Loading soils up with fertilizer serves crops a free, all-you-can-eat buffet. Their response? They cut back on making and delivering phytochemical-rich exudates into the rhizosphere to attract and retain nitrogen-fixing soil bacteria. So as green bodies pig out on nitrogen, growing bigger leaves and fatter fruits, we pay a hidden price in plant foods with fewer phytochemicals.

Growing Health

Resorting to pesticides and relying on chemical fertilizers is about the only way farmers and gardeners can compensate for plants that dial back on their phytochemical production and curtail relationships with their underground farmhands. This effect takes the rhizosphere from a bustling biological bazaar to a ghost town, unraveling the defensive strategy of crops—and nature's grand plan that has a half-billion-year track record of keeping green bodies humming along.

A clear example of undermining the natural defenses of plants comes from research on orchard crops. Applying a lot of nitrogen fertilizer to apple trees spurs growth but reduces levels of phenolic phytochemicals in young leaves. This, in turn, makes the trees more susceptible to apple scab. Plants make phenolic compounds in proportion to the relative availability of carbon and nitrogen. And in a true case of doing more with less, plants faced with limited nutrients generally increase their production of phenolic compounds. This, in large part, is why nutrient-rich environments like nitrogen-fertilized soils can lead to decreased phytochemical production.

All told, abundant evidence shows that agrochemicals influence soil life in ways that affect crop resilience and health. Consider the global demand for pesticides—it rose alongside fertilizer use in the twentieth century. Increased reliance on synthetic nitrogen to boost crop growth allowed tilled monocultures to become the norm. The resulting soil degradation and loss of diversity above- and belowground undermined crop health. In hindsight, reliance on chemically fertilized, tilled monocultures cemented the need for pesticides.

All this began decades ago, well before science grasped the role of microbial soil ecology in plant health. For insight we can look to studies on the effects of herbicides like glyphosate on soil life and plant health. What, if anything, does the most commonly used herbicide in the world do to crop health and nutrient density?

As early as the 1980s, studies documented that glyphosate applications made plants more vulnerable to fungal root rot. Subsequent studies confirmed this finding and identified a likely mechanism. Glyphosate use alters rhizosphere microbiota in ways that compromise the ability of crops to suppress root-colonizing pathogens. For example, one study documented that applying glyphosate to glyphosate-resistant soybeans reduced populations of beneficial soil fungi and bacteria, promoted pathogenic fungi, and reduced plant shoot and root biomass enough to prove detrimental to plant growth and productivity.

Studies that show a tendency for fungal pathogens to increase in proportion to applications suggest glyphosate use heralds the appearance of more pathogens more often. It reportedly increases root disease in wheat when used in no-till fields. A striking example comes from an extensive study of wheat and barley on Canadian prairies that demonstrated prior glyphosate application was the most important agronomic factor associated with the development of pathogenic *Fusarium* fungi. Applying glyphosate before planting to "burn down" weeds also has been associated with higher incidence of root rot in both wheat and barley.

A particularly compelling, ten-year study from 1997 to 2007 at the University of Missouri found that glyphosate applications resulted in greater root colonization by *Fusarium*. Field trials at two research stations and six farms found pathogenic fungal root colonization was two to five times higher for soybeans treated with glyphosate than for those receiving none. In addition, glyphosate applications reduced populations of beneficial bacteria that help combat *Fusarium*.

These were not isolated findings. A 2009 review in the *European Journal of Agronomy* lists a wide range of crop diseases shown to increase after glyphosate use. One review of the effects of glyphosate even suggested that the popular herbicide could present a long-term threat to agricultural stability.

In addition to increasing vulnerability to disease, glyphosate has been reported to influence the availability of mineral micronutrients.

Studies have shown glyphosate to substantially reduce uptake of copper and zinc, which are essential for plant defense, as well as manganese and iron in sunflowers and calcium and magnesium in soybean seedlings. That even low residual levels of glyphosate can reduce root mineral uptake suggests that glyphosate drift or residues can affect nontarget plants.

Glyphosate also reportedly stimulates soil microbes that bind up mineral nutrients in stable oxides, making them unavailable to crops. The result is the potential for such nutrient deficiencies to render plants vulnerable to pests and disease.

Glyphosate also interferes with the ability of nitrogen-fixing bacteria to colonize roots. And this limits the ability of the plant to acquire nitrogen, which increases the need for fertilizer. Still, it's fair to say we lack a full understanding of everything that glyphosate can do to crops and soil life.

Whether crops are genetically engineered or not, relying on chemical fertilizers increases demand for pesticides, while pesticides in return increase demand for fertilizers. Why has this aspect of how we farm been overlooked or ignored for so long?

Adopting yield as the ultimate metric sidelined concerns over potential negative effects on plant or soil health. Consider, for example, the interpretation of a 2012 review that reported conflicting conclusions in studies about whether glyphosate use affects mineral density or crop disease in glyphosate-resistant crops. The authors relied on the observation of unchanged trends in yield following adoption of glyphosate-resistant soybeans, cotton, and corn to dismiss potential adverse effects of glyphosate on crop mineral uptake or disease susceptibility. In other words, they confused crop growth with crop health.

Compromised mineral uptake is one way glyphosate can affect crop health. Minerals are essential components of enzymes that regulate botanical mechanisms central to stress and disease resistance. When glyphosate binds up manganese, it deprives plants of a cofactor that activates almost three dozen enzymes involved in the synthesis of phytochemicals. Manganese-deficient plants produce less lignin and flavonoids, rendering them more vulnerable to pathogens and disease.

Another way that glyphosate can reduce disease resistance relates to phytochemicals. In the 1980s, soon after its commercial adoption, studies documented that disrupted phytochemical synthesis increased the severity of fungal root rot in glyphosate-treated beans and weeds. A decade later, researchers identified 24 diseases in 14 crops that had increased under glyphosate-based weed control programs. The authors of that review went so far as to identify glyphosate as a significant factor in the reemergence of crop diseases previously considered under control. They concluded that the herbicide compromised the ability of plants to defend against soilborne pathogens.

The issue of lessening disease resistance goes beyond glyphosate, as the same effect is associated with other herbicides as well. A 2014 review concluded it was generally accepted that herbicides weakened plant defenses enough to increase crop vulnerability to pathogens.

Unconventional Wisdom

Chief differences in the health of crops grown organically or conventionally are linked to the effect of levels of soil organic matter on soil life. Crops in organic farming systems generally have less severe root disease and pest problems than conventional systems. Along these lines, a 1995 UC Davis review attributed higher disease suppression in soils farmed organically to a higher abundance of mycorrhizal fungi compared to levels in conventionally managed soils.

Numerous studies have shown that organic farming increases levels of soil organic matter relative to conventional farming. Likewise, organic fields consistently have greater microbial biomass, diversity, and activity. Such differences take time to manifest, however, as differences in soil health typically do not become apparent for several years after switching from conventional to organic practices. And while plant-available nitrogen is usually lower under organic practices, more soil organic matter enhances microbial cycling of potentially available nitrogen and soil minerals.

Assessments of the effects of farming practices on food quality are usually cast in terms of whether organically grown food is better than conventionally grown food. And while ongoing controversy surrounds this issue, studies have found some interesting patterns. You just have to dig beyond the headlines to sort them out.

Health-conscious urbanites, concerned parents, and dedicated proponents of organic farming generally take it on faith that organic food is better for us. Defenders of conventional agriculture tend to paint such talk as antiscientific blather. In recent decades, supporters of these incompatible viewpoints argued past one another as organic farming expanded into the fastest-growing agricultural sector.

Along the way, some studies found significant differences in the nutritional value of foods. Others reported none. Naturally, this fueled impassioned arguments about any health benefits for organic over conventionally grown foods. Every other year or so it seems a new headline-spawning study fires another volley in the running argument over whether organic foods are more nutritious.

Understanding the implications of such studies requires diving into specific differences that have and have not been found, a consideration of study design, and the potential for the influence of confounding factors to mask differences. Other things that matter a lot are the particular crop variety or cultivar, the geologic nature of the soils, and even the weather as a crop grows. And as we'll see, evaluating the effects of farming practices on nutrition also hinges on what one defines as nutrients. So what can we learn through all the variability, politicized advocacy, and industry spin? Are we just faced with a horribly complex issue with no simple answers, or are we perhaps looking at it through the wrong lens?

Central to sorting through these ongoing arguments is how we think and talk about nutrition and nutrients. When reviewing the recent scientific literature on this topic, it becomes apparent that investigators tend to focus on aspects of nutrition central to growth, like carbohydrates and protein, rather than micronutrients and phytochemicals central to lifelong health. Because media reports rarely consider how researchers defined nutrients, it's worth digging a little deeper into what various studies found.

One of the first published comparisons of the nutritive value of crops fertilized with manure versus chemical fertilizers comes from a 1974 German study. Based on 12 years of data on various crops fertilized with farmyard manure, compost, or the standard NPK trio, those grown with farmyard manure or compost had more protein, vitamin C, phosphorus, potassium, and calcium, as well as significantly less nitrate and sodium than those grown with NPK. Spinach had an astounding 77 percent more iron. Overall, the study concluded that organic practices produced more-nutritious, lower-yielding crops. These findings helped frame the narrative of organic farming favoring quality over quantity.

Two years later, a comparison of tomatoes, potatoes, peppers, lettuces, onions, and peas grown with organic or "commercial" fertilizers found no difference in major elements (nitrogen, phosphorus, potassium, calcium, and magnesium). Such findings set up the narrative that there are no differences in the nutritional quality of conventional and organic foods. In this study, however, the influence of soil health under the two systems was not a factor as both crops were grown in the same soil. This study design left differences in micronutrients and phytochemicals unaddressed, and yet dietitians and nutritionists repeatedly remind us all to eat foods rich in both.

Another comparison, a 1993 study published in the *Journal of Applied Nutrition*, reported significant differences in the mineral content of conventional and organic foods purchased from grocery stores in suburban Chicago. Over a two-year period, organically grown wheat, corn, potatoes, apples, and pears averaged 60 to 125 percent more iron, zinc, calcium, phosphorus, magnesium, and potassium than their conventionally grown counterparts. In other words, the organic foods had roughly half again to more than twice the mineral content of their conventional counterparts. Since then, high-profile studies that compared conventional and organic foods reported a wide range of results and contradictory conclusions.

A popular approach to dealing with this complexity has been to look for trends among large numbers of studies. However, a key problem with these meta-studies—studies of studies—is that all kinds of practices get lumped together under the labels of conventional and organic. Additional complications include the effects of soil type, climate, and different experimental designs between studies—like controlled field experiments versus farm and retail surveys. Inclusion of conventional no-till farms with organic matter–rich soil and organic farms with frequently plowed, organic matter–depleted soil further complicates assessing differences given the wide range of practices and soil health on both conventional and organic farms.

But there is an odd upside to the statistical blender of a mix of conditions and external factors that make it more difficult to detect real differences. Those detected in meta-studies are likely real, whereas studies finding none might mask underlying differences.

And simply basing nutritional comparisons on the organic versus conventional distinction can blur differences due to soil health. Whether organic and conventional farms degrade or build soil health depends on how their specific practices affect soil life. Occasional sparing use of judicious amounts of chemical fertilizer may be less harmful to soil health than routine plowing. Similarly, in some circumstances occasional tillage may not compromise soil health.

Studies that directly compare farming methods on crops grown in the same soil don't face the problem of mixing comparisons between regions or different land-use histories. And legacies of past farming practices can matter as well. Published studies comparing conventional and organic crops generally fail to take into account how long the fields were worked using organic methods.

Nonetheless, some patterns do emerge.

A 2001 review of 41 studies reported that organic crops on average had 20 to 30 percent more vitamin C, iron, and magnesium. Though the differences varied by crop, overall levels of mineral micronutrients were consistently higher in organic crops, a difference attributed to the influence of soil life. Conversely, levels of heavy metals were higher in conventional crops. The study also reported that nitrate levels were significantly higher in conventional foods for almost three-quarters of 176 cases where nitrate levels were compared. While organic foods averaged 15 percent less nitrate, the nitrate levels in conventional foods ranged up to eight times higher. A number of other studies also found conventional foods contained more nitrates than organic foods, with a 1997 review concluding that conventionally grown vegetables had "far higher nitrate content than organically produced or fertilized vegetables."*

Another 2001 meta-study, published in *Critical Reviews in Food Science and Nutrition*, found higher levels of micronutrients in organic vegetables and legumes than in conventional ones for comparisons that controlled for crop variety (cultivar) and soil conditions. Overall differences were modest, however, with up to 10 percent differences in beta-carotene, vitamin C, boron, copper, and zinc. But for studies that screened for the same cultivar and soil conditions, the micronutrient contents of organic foods ranged up to almost half again higher than conventional produce. The message here was that carefully controlled comparisons revealed large differences.

Perhaps the most widely publicized meta-study on differences

---

* Woese, K., D. Lang, C. Boess, and K. W. Bogl. 1997. A comparison of organically and conventionally grown foods—results of a review of the relevant literature. *Journal of the Science of Food and Agriculture* 74, p. 290.

between organic and conventional food was an ambitious 2012 paper with the imprimatur of the Stanford Medical School. The authors reviewed 223 studies, reporting that conventional produce consistently had significantly higher levels of pesticide residues and antibiotic-resistant bacteria and that organic produce had higher levels of phenols. They found no significant differences for potassium, calcium, magnesium, and iron nor for vitamins A and C. Headlines trumpeted the conclusion that organic food was no more nutritious than conventional food despite the latter containing more pesticides and fewer health-promoting phytochemicals.

Just two years later, in 2014, an even more comprehensive study of such differences in the *British Journal of Nutrition* expanded on the very same data set to analyze 343 peer-reviewed publications. This even more exhaustive study reported significantly greater levels of antioxidants in organic crops and significantly higher concentrations of pesticide residue and the toxic metal cadmium in conventionally grown crops. In particular, it found that organically grown foods contained higher levels of a number of phytochemicals, estimating that people would take in 20 to 60 percent more phytochemicals if they ate organic rather than conventional foods. So you'd have to consume about twice the conventional produce—pesticides and all—to get the same amount of phytochemicals.

This was not exactly a new finding. Yet another 2001 review, this one in the *Journal of the Science of Food and Agriculture*, came to the conclusion that phytochemicals were 10 to 50 percent higher in organic vegetables than in conventional vegetables. Similar findings came from a broader follow-up meta-analysis a decade later. The differences between organic and conventional fruits and vegetables were interpreted as reflecting organic crops having higher exposure to soil microbiota that spur plants to increase production of defense-related phytochemicals.

A more recent meta-analysis, published in 2018, looked only at studies from after 2000 due to concerns about earlier experimental designs and changing measurement techniques. It too found that crops from organic systems had higher concentrations of vitamins and phytochemicals, particularly antioxidants, carotenoids, flavonoids, and phenolic compounds. Not surprisingly, it also found conventionally grown fruits and vegetables consistently had higher pesticide levels than organic crops.

Phytochemical levels also vary substantially between different varieties of the same crop and even between individual plants of the same crop. In addition, a three-year UC Davis study of the levels of phytochemicals in tomatoes found significant variation from one year to the next, reflecting the effects of weather. Despite such influences, clear patterns run through direct comparisons.

A 1991 French study of two dozen pairs of farms growing the same crop varieties in the same type of soil found that relative to their conventional counterparts, organic carrots had more beta-carotene and vitamin B_1 and organic celery root had more vitamin C and less than half the nitrate. Likewise, a 2004 Australian comparison of wheat grown under organic and conventional management found no difference in most macronutrients but a third more to half more zinc, as well as more copper, in organic wheat. Zinc levels tracked with colonization by mycorrhizal fungi, which conventional management depressed.

And a 2010 study of strawberries from 13 paired organic and conventional farms in California, each with side-by-side plots, found significantly higher levels of polyphenols, vitamin C, and antioxidants in the organic ones. The organically farmed soils had higher carbon and nitrogen levels, greater microbial biomass and genetic diversity, and significantly higher levels of plant-available iron and zinc. European studies of potato and plum cultivation found that organic farming practices resulted in more phytochemicals (polyphenols and anthocyanins) and better taste, but that the enhanced quality was only for certain varieties better suited for organic practices.

Well-controlled, long-term studies of phytochemical levels in organic and conventionally grown crops are rare because of all the factors that affect crop growth and health. But in 2007, researchers at UC Davis published such a study. Over a ten-year period, they documented that mean levels of two phytochemicals known to benefit human health, quercetin and kaempferol, were 79 percent and 97 percent higher, respectively, in organically grown tomatoes than conventionally grown tomatoes. While the flavonoid levels in the conventional tomatoes stayed relatively constant over the decade-long study, those in the organically grown tomatoes increased over time as the organic matter content of the soil increased. As soil health and microbial activity improved, so did the phytochemical content of the tomatoes.

Organic management and fertilization practices have been found to increase total phenolics in eggplant, pears, peaches, corn, apples, marionberries, and blueberries. A 2005 review in the *European Journal of Nutrition* found that agronomic practices could double the content of individual phytochemicals in radishes and increase them tenfold in broccoli and cauliflower. Conversely, nitrogen fertilizer use can decrease the amount of glucosinolates in broccoli by as much as 70 percent.

A particularly well-controlled Japanese experiment in 2001 compared the antioxidative and antimutagenic properties and flavonoid content of the same varieties of five green vegetables (Chinese cabbage, spinach, Welsh onion, green peppers, and qing-gen-cai) harvested on the same day from adjacent organic and conventional farms. The organic vegetables had 20 to 120 percent higher antioxidative activities and greater ability to suppress mutation in cell cultures. Flavonoid content was much higher in organic than in conventional vegetables, with one type, quercetin, occurring at levels up to ten times higher. In short, the preponderance of results from a wide range of studies shows that cultivation practices greatly affect phytochemical content.

Perhaps the best example to date of how farming practices influence the nutritional quality of food comes from the farm systems trials at the Rodale Institute in Kutztown, Pennsylvania. Since 1981 the institute has conducted an ongoing comparison of organic and conventional practices on full-scale field plots. Growing the same crops in side-by-side plots eliminated most of the effects that complicate meta-study analyses of dissimilar experiments.

In 2003, 22 years later, the soil organic matter and nitrogen levels had increased on the organic plots but not on the conventional plots. Oats from the organic system had roughly a third more minerals overall, ranging from 7 percent more potassium to 74 percent more boron. Iron was 23 percent higher and zinc 40 percent higher. Vegetables grown in the plots in 2005 also had large differences in total antioxidant and vitamin C levels. Organically grown tomatoes and jalapeño peppers, respectively, had 36 percent and 18 percent more vitamin C. Organic carrots had 29 percent higher total antioxidant levels. These differences reflect levels of crop uptake from the same soil in the same year.

What was behind the difference? Soil health. A little over a decade into this experiment, in 1994, an assessment of soil biology was conducted. Measured by plant-available nitrogen and soil respiration rates, the conventionally managed plots had lower biological activity than the cover-cropped organic soils, which increased in carbon over the course of the field trials. Healthier soils with more soil life were producing more nutrient-rich crops.

What's the bottom line from all these studies? Conventional crops often look better in terms of yield. But by the time they land on our plates, they contain more of the elements we add as fertilizer to a person's diet, most notably nitrogen (which few Americans are short on). They also have more things that shouldn't be in food at all—pesticides and heavy metals. But the most striking, usually overlooked result of these studies is that they consistently find significantly more phytochemicals in organic foods, along with more micronutrients (vitamins and minerals) important for health. In other words, much of the confusion around whether farming practices affect the nutritional value of food stems from what gets analyzed—macronutrients, micronutrients, or phytochemicals—and how we define nutrients.

Finally, any nutritional comparison must consider yields, as well as nutritional quality. A 1990 Israeli study published in the journal *Agriculture, Ecosystems and Environment* examined 205 comparisons of crop yields from studies that involved comparative observations, long-term replicated field plots, and Balfour's Haughley experiment. The study found that across 26 crops, organic yields averaged almost 10 percent lower than conventional yields. This is the standard wisdom one typically encounters in discussions surrounding this issue. But after digging a little deeper into the paper, one finds that organic crop yields equaled or exceeded conventional yields in half of the 30 direct yield comparisons. Similarly, in the long-term field experiments reviewed, only half of the comparisons found statistically significant differences between organic and conventional yields. A quarter of the harvests analyzed showed that organic crops outproduce conventional ones. The key message here is that in many cases organic yields were comparable to or better than conventional yields.

More recently, a 2015 study that reviewed previous studies comparing yields of organic and conventional farming systems found that while organic yields were roughly 20 percent lower overall, diversified organic crop rotations reduced the yield gap to less than 10 percent. The authors concluded that appropriate research investments in organic cropping systems could further reduce if not eliminate the yield gap. The most recent review, in 2018, noted that yield differences between organic and conventional systems declined over time. It took only several years of organic production to significantly reduce or even eliminate the yield gap. So with the right mix of practices farmers do not have to choose between quantity and quality. Perhaps a better way to frame the question for any farming system is to assess whether practices improve or degrade soil health and thereby help or hinder harvesting both quantity and quality.

And don't forget profitability. A global analysis found that organic farms were a quarter to a third more profitable than corresponding conventional systems. While prices for organic crops are generally higher than conventional counterparts, this global study found that even assuming lower yields from organic farms, prices need only be 5 to 7 percent greater to match the economic performance of conventional cropping. While organic foods do command a premium in the market today, their price differential need not be great.

An important, under-recognized factor in many meta-studies concerns organic farms that were previously farmed for decades using conventional practices that left the soil depleted of organic matter. Soil, like a person, does not recover from harm and trauma overnight. Conventional crop varieties grown without synthetic fertilizers or pesticides on degraded fields produce low yields using organic methods. A fairer comparison would involve crops bred for organic practices grown in fields with healthy, fertile soil. It is telling that the yield advantage of conventional farming does not seem to hold on soils rich in organic matter.

All in all there is a need to reframe nutritional and agricultural research and policy around the effects of farming practices on soil health. Degraded soil undermines both phytochemical production and micronutrient acquisition, leaving crops more vulnerable to pests and pathogens. This, in turn, leads to greater demand for pesticides that end up in our food and impact the health of farmers, farmworkers, and consumers. And plants on a soil health diet don't just contain fewer pesticide residues. They often contain more phytochemicals and micronutrients.

Just how much does nutrient density improve when food grows in healthy, fertile soil? We didn't find studies that addressed this question directly, and so we decided to see for ourselves. Perhaps there really isn't a devilish trade-off between abundance and quality after all.

7 BIG GREEN THUMBS

To pay attention, this is our endless and proper work.

—MARY OLIVER

How much of a difference does healthy, fertile soil make to the nutritional quality of food? Coming face to face with the pale, wretched dirt beneath our old-growth lawn when we started a garden motivated us to dive into this question. At first, robins didn't even bother prospecting for worms in our yard: there were none. But over the course of a decade a steady diet of wood chips and organic matter mulches brought the soil back to life. And as the carbon content of our vegetable beds rose from 1 to 10 percent, we became increasingly curious about the nutrient makeup of our backyard crops.

So we submitted a sample of kale to a lab and learned it had far higher levels of calcium, zinc, and folic acid (vitamin B_9) than the USDA nutritional standards for conventionally grown kale. We also learned our kale contained 31 parts per million of sulforaphane, a cancer-fighting phytochemical. Unfortunately, there are no nutritional standards set for sulforaphane—or other phytochemicals—despite, as we'll see, an abundance of evidence that they help prevent a number of chronic diseases.

While it's one thing to grow nutrient-dense vegetables in a small backyard plot, a vegetable farm faces more daunting challenges. The greatest one is growing vegetables without plowing the beds or fields—or so a number of organic farmers and conventionally minded agronomy professors told us. In their opinion combining no-till and organic practices simply would not work for vegetables: they argued that without tillage, weeds would overrun crops. But we'd heard of vegetable farmers in Connecticut and California who had devised no-till methods. Naturally, we wanted to meet the farmers and learn about their methods—and find out how their no-till vegetables compared to conventional produce.

Tobacco Road

It's an overcast day in late June as we drive through southern New England passing horse stables, rural homes, fancy estates, and stands of unbroken forest. The greenery beneath a pewter-colored sky reminds us of early summer in Seattle. But every time we open the car window the similarity ends as we breathe in heat and humidity.

Our phone's GPS informs us we have arrived at our destination, Tobacco Road Farm, and we pull into a long gravel driveway. We're not so sure. No fields are visible, just seemingly haphazard piles of chopped wood, leaf mulch, straw, rock dust, and things in various stages of decay beside a weathered hoop house.

We know we're in the right place once Bryan O'Hara emerges from his cedar-shingled, cabin-like house. Bryan wears a beaming smile and a green T-shirt, blue jean overalls, and Birkenstocks, with a hearty beard and brown shoulder-length hair flecked in gray. His wife, Anita, and daughter, Clara, walk up a few minutes later and greet us.

The smell of fresh garlic reaches us long before we spot the bed filled with a ready-to-harvest crop. Bryan began farming at Tobacco Road Farm in 1990. It had poor, acidic soil beneath a big field of wild blueberry and poison ivy. He's eager to show us how much it has changed.

Over the past decade Bryan and Anita transformed their soil, land, and farm. After meeting at the co-op in town they discovered a shared desire to grow food without pesticides and other chemicals. They began developing methods to build and maintain soil health and saw the connection between soil fertility and crop health play out on their farm. We could see it too, as there wasn't a sickly or diseased crop among the many covering and spilling out of the beds. Originally their soil had about 3 percent organic matter and a pH of about 4, acidic enough for aluminum toxicity to be a real issue. Now the pH is close to neutral, and their soil organic matter is up to 11 percent, well above that of the precolonial soil.

At one time there were lots of sheep and wool mills in the region. But the area was virtually abandoned in the late 1800s. Shortly thereafter moonshiners moved in near where the O'Haras' farm now stands. Bryan gave the farm the name of the road from that time.

With about 150 years of regrowth the surrounding forest once again sports big trees. But they are declining again. "We've already lost the chestnut, but ash, spruce, birch are all dying out," Bryan laments. He can't tap his sugar maples anymore; they don't produce like they used to. He thinks these changes reflect a broader imbalance in nature, and about a decade ago he and Anita felt they had to up their game. At the time, problems like pests and weeds were becoming bigger challenges. In his mind the situation called for working against environmental collapse, for greater attention to soil health.

As we walk their three-acre farm, Bryan tells us, "When we got into vegetable farming 30 years ago, the older farmers were skeptical. A lot of that generation died early of chemicals. Their kids are not interested in doing the same thing." Times are changing though, and the local agricultural extension service now offers advice on organic practices and soil health.

Bryan also relays that he has seen a big decline in insects over the past decade. He had to abandon honey production due to repeated bee losses, and there are also fewer fireflies, katydids, crickets, and even mosquitoes around the farm. Belowground, though, it's going better. The soil supports plenty of worms, and they attract a lot of birds. As if on cue, an eye-catching pair of red wings flash past.

We head over to the mulch piles, and Bryan launches into telling us more about them. "What's going on here is mixing." He makes mulch or compost from four materials: straw, wood chips, leaves, and rock powder. The first three contain a lot of carbon, he explains, and so when he uses these materials for compost, he adds a nitrogen source (manure) to balance out all the carbon. The final ingredient is powdered granitic rock as he finds it enhances crop growth. Bryan and Anita use a lot of organic matter in their operation, more than 50 tons a year.

Through his composting techniques Bryan aims to keep the compost temperature under 120°F. His twist on standard compost methods is to let microbes partially digest the organic matter he uses to make the compost. Once applied to the vegetable beds, soil life finishes the job, slowly dripping nutrients into the soil.

Doesn't it cost a lot to acquire that much organic matter? Bryan points out that vegetables are a fast-growing, high-output crop—the most profitable crop per acre you could grow legally in Connecticut at the time of our visit. An acre can yield $100,000 or more of produce each year, far more than can be harvested off an acre of grain. He says growers need to understand that a high-value vegetable crop allows for investing in fertility-maintaining compost and organic matter. In fact, Bryan concludes, that's the key to successful no-till vegetable production.

Animal manure, especially from cattle, is the traditional manure of choice for vegetable growing. But the practice stopped when agrochemicals became conventional. Yet Bryan remains adamant: "You have to add organic matter. I've never seen failure to apply fertility materials work out well for vegetable farmers." Bryan and Anita get manure from grass-fed, pastured animals that "never need to see veterinarians." Manure from conventional dairies or feedlots won't do as it contains antibiotics likely to harm soil life. They consider state-mandated manure lagoons on modern dairies misguided because the manure doesn't get composted and put back onto the land.

Bryan leads us out to fields where rows of crops stand healthy and vibrant in multiple hues of green beneath the overcast sky. Scattered glimpses of bare ground reveal rich, black earth peeking through the remains of prior crops and decaying wood chips. The soil is spongy and easy to scoop up by hand. All the organic matter gives it a crumbly, granular texture, and the worms seem to love it, wriggling around, amped up on the good life.

When Bryan and Anita decided to go no-till, the biggest challenges were controlling weeds and managing fertility. They found that one overarching practice works really well—grow something everywhere all the time.

They keep their beds busy growing year-round all across the farm. The farm is divided up into multiple long rows of different crops or pairings of companion crops sharing the same bed. In the first bed we come to, lettuce and tomatoes are intercropped. He and Anita don't dig at all; instead they plant each new crop into the mulched remains of the just-harvested crop. Arugula, turnips, carrots, and more, they grow over 50 different crops in this manner.

Each crop gets a narrow section of the field, and every year each crop shifts over one section. Because of the large variety of crops Bryan and Anita grow, a long stretch of time passes before the same crops ever get planted again in the same place.

The garlic we smelled as we arrived is just about ready for harvest.

Bryan pulls a fat bulb out of the ground. It's huge and fragrant, sweetly reeking of phytochemicals. Anita tosses it into the basket she's carrying.

There are a few short weeds growing in the bed, but Bryan considers it no real concern. His crop is shading out the weeds, and any remaining after harvest will be mowed down before going to seed and then mulched in place before the next crop is planted. Here is the secret to weed control on their farm: weeds never get a chance to take off, much less take over. The crops get a head start and outcompete them. This turns any weeds into green manure that acts as fertilizer for the next cash crop. In some plots they leave clover to take advantage of its nitrogen-fixing abilities, while other plants support pollinators. If weeds aren't harming a crop, they can stay—until everything gets mowed before the next crop is planted. Bryan and Anita are making money recycling weeds instead of spending time and money trying to kill them.

They work three fields, the main one on the farm and two on nearby rented land. Enough rain falls (45 inches a year) that they never need to irrigate. All the rain soaks into the lush, spongy ground. Hardly any runs off, even in intense downpours. After they went to a no-till system, they also got rid of their drip irrigation system. It paid off. The combination of no irrigation and great weed control saved a lot on labor.

For Bryan, everything relates back to the soil. "It's what farmers have the biggest influence on and the primary element you can influence." So he's not afraid to manipulate it, constantly experimenting, observing, and tinkering.

Doing so, he's landed on a winning recipe. After harvesting a crop and mowing down any residual weeds or crop parts close to the ground, he covers the bed with clear plastic. After a day or two in the hot sun any stragglers die. He rolls up the plastic sheeting to use again and then layers about a half inch of the partially decomposed compost he makes on top of the soil, sometimes adding a fungal inoculant as well. After broadcasting seeds over the bed surface, he finds that germination is better if he increases seed contact with the soil. So he has attached lengths of chain rings arranged like the tines of a fork on a large rake-like tool. One time back and forth across the bed and the final touch—a bit of mulch—and he's done planting the bed.

Bryan seeks to drive as much organic matter decomposition and photosynthesis as possible "to keep the entire ground surface covered by plants or mulch at all times." True to his words, we didn't see any bare fields on the farm. The textures, colors, and shapes of all the crops formed a dazzling, carpet-like mosaic. No field lies fallow on this farm. They all thrive, bursting with life.

As Bryan and Anita stopped tilling, they also started feeding their soil, adapting methods known as Korean Natural Farming to their farm. This involves mixing a bit of the nearby forest soil rich with mycorrhizal fungi into the compost and mulch. The idea is to jump-start fungal activity. The combination of no disturbance and inoculated organic matter worked really well. The soil is now shot through with fungal mycelium, and mushrooms routinely pop up in the fields. Belowground they have far more worms than they used to, and the organic matter content of their soil rose to more than three times higher than the soil at the conventional farm next door.

Running throughout our conversation with Bryan was his insistence on constant observation of the farm's conditions—both natural and agricultural. He put it to us succinctly: "Everything in nature is communicating." He tells us how the forest and ground-hugging, broad-leafed burdock encircling the fields provide habitat for snakes, frogs, birds, and beneficial insects that eat would-be pests looking to dine on their veggies.

As we stand around a densely planted bed of lacinato kale, Bryan shows us what he means. He turns over several of the dark green, rugose leaves and points out a small cache of moth eggs on the underside, future pests waiting to hatch and chow down on his crop. He's unperturbed; he has seen the eggs before and knows what they are.

Better yet, he knows what will eat the caterpillars soon to emerge from the eggs.

The fields attract parasitoid wasps that will lay eggs of their own in the caterpillars, turning them into a ready meal for baby wasps before the former ever get the chance to damage his kale. As the young wasps mature into adults, they switch their diet from caterpillars to nectar from cabbage family plants and overwinter in a cocoon beneath their leaves. Providing overwintering habitat for the adult wasps ensures a new batch of wasps is ready in the spring before any voracious caterpillars attack his crops. So Bryan and Anita always let some cabbage family crops keep growing beyond fall.

As we walk the farm, Anita keeps filling her basket with salad ingredients pulled up along the way. By the time we complete the tour and get back to their house, she has collected the makings for the freshest lunch imaginable—just-picked fennel, two kinds of peas, celery, and several varieties of lettuces, served up with cheeses and a touch of super-fragrant garlic. It was a delicious demonstration that good soil grows great-tasting food.

When Bryan and Anita jumped into starting a vegetable farm, the demand for healthy food was growing. Their timing was fortuitous. Among the early customers who began showing up at the farmers' markets where they sold their produce were people with cancer, multiple sclerosis, Parkinson's, or digestive disorders. They wanted pesticide-free food or, as Bryan put it, "vegetables that weren't poisonous and didn't taste like crap."

Now Bryan and Anita sell most of what they grow wholesale to the co-op and local restaurants in town and the rest at farmers' markets on weekends. So one way or another they always sell their crops. "We've never done any marketing whatsoever. The vegetables sell themselves." When Bryan looked into renting even more fields and expanding, the extra costs meant they could "work harder to be in the same place." That didn't make any sense, so they stayed small and continued to thrive.

At the same time, Bryan realizes he's rowing upstream. With farmers presently paid on volume rather than quality, he sees why they use excessive nitrogen fertilizers and pesticides: it's a recipe for volume.

Bryan's recipe and objectives are different. He merges old-school methods with modern science to focus on high-quality crops that he believes deliver produce far superior to conventional vegetables. He judges quality based on three characteristics: storage (how long a vegetable stores, especially root crops), flavor (how good it tastes), and color (vibrancy). So how do you get quality in the crop? Not tilling the soil, not overfertilizing, keeping soil biology active, and providing habitat for insect predators that chow down on pests.

Interestingly, while the soil on Tobacco Road Farm is naturally low in copper and zinc, the tomatoes are high in copper, and the spinach tends to have high amounts of zinc relative to what's considered sufficient. Tests using strong acid that dissolve rock particles reveal plenty of other minerals in the soil as well. His kale is quite high in molybdenum, and his lettuce is rich in iron, as is Bryan himself.

What is moving those mineral micronutrients out of the ground and loading them into Bryan and his crops? Soil life. Bryan emphasizes that plants need mineral elements to make many of the enzymes crucial for their health, and he further explains that when practices hamper enzyme production, insects and disease strike because the plant can't make the compounds it needs to defend itself. Bright pigments in the color of their flowers are among the telltale signs that his crops are getting adequate deliveries from his underground farmhands.

Bryan and Anita are no starry-eyed farmers. They have toiled and sweated to turn the soil around and make a good living, enough to buy Tobacco Road Farm. But they fear others won't get the chance. Banks today favor lending for large, expensive houses, which makes it hard for a property to serve as a working farm. They share a vision for the future of agriculture where practices on each farm are uniquely tailored to the soil and the natural setting of the land. And it turns out that Bryan was finalizing a book when we visited. *No-Till Intensive Vegetable Culture* shares his growing methods and accumulated knowledge with what he hopes will be a new generation of farmers—those striving to bring life back to their land and their soil back to life.

Right before we head off, a chorus of birds visiting the farm provides a robust background serenade. It's a fitting close to the day.

Singing Frogs

What Bryan and Anita have done on their Connecticut farm has striking parallels to the next stop on our journey—Singing Frogs Farm, a small-scale vegetable farm nestled in the wine country north of San Francisco. The conditions are hot and dry compared to the verdant, humid landscape around the O'Haras' farm. Nevertheless, Paul and Elizabeth Kaiser coax an equally dazzling display of veggies from the earth without tillage or agrochemicals.

We look out the window of the plane as it descends to the small Sonoma County airport. The northern California landscape is familiar, reminding us of our time as graduate students in the Bay Area. Expansive stands of native oaks mix with redwoods between golden-brown, grass-covered hills. It's July, well past the rainy season, and the land is parched. Blocks of grapevines come into view, some laid out in tidy rows, others following the natural contours of the land. But on most vineyards the soil lies naked and pale between the rows.

The plane lands, and we set out to meet Ray Archuleta. Known in the farming community as "Ray the soil guy," he is a passionate and boat-rocking scientist recently retired from the Natural Resources Conservation Service. He pushed the agency to move beyond its historical roots in erosion control and helped spearhead a new emphasis on working with farmers to build soil health.

Ray would be helping us compare the nutrient density of a crop from Singing Frogs to that of a paired farm with very different soil health. While we'd handle the crop samples, Ray was in charge of soil samples.

The next morning we leave the small downtown area of Sebastopol, and in no time we're on a single-lane road with sharp turns running past rural houses and small vineyards. We come to the end of the road past a trailer emblazoned with "Compost Club" and park the car. A pair of goats boldly stare at us with golden eyes from their enclosure bearing a frog crossing sign. We stare back at them as Paul and Elizabeth walk up and greet us.

Their not-quite-nine-acre property is shaped like a slice of pizza, with the crust at the uphill end where their house sits. The property narrows downslope to ponds at the bottom of the farm and distant views of vineyards that cover the slopes on the opposite side of the valley. They keep two-and-a-quarter acres in vegetable production, and their barns, hedgerows, and compost piles take up another half acre. The remaining acreage provides space for mostly wild lands, the ponds, and a spot for their house.

When the Kaisers got the place in 2007, it had been an organic hobby farm for 12 years. There were almost no trees and only two bushes. It was a perfect setting for them. They had field experience in tropical agroforestry from their time in the Peace Corps in West Africa. Now several thousand native perennials, fruiting shrubs, and trees grace the farm. Unlike a lot of areas in California, the Kaisers' land has water due to the series of spring-fed ponds throughout the farm. They call it Singing Frogs Farm for good reason.

In his mid-40s now, Paul grew up in Menlo Park near the original Grateful Dead house wanting to be a park ranger and teach people about nature. Now he's doing that, only on a bird-friendly farm. The day we visit he's wearing a farmer's version of a ranger outfit—a black Australian cowboy hat, purple-and-black plaid shirt, and brown work pants with lots of pockets. Some nearby parks and nature reserves don't have as much avian life as Singing Frogs, according to Paul. They attribute the influx of wildlife to planting all those trees and shrubs. But the real key to Paul and Elizabeth's farming success is that they've restored life belowground.

When they got the farm, the previous owner helped them get started. How? By plowing everything up. Although Elizabeth now farms full time, she initially worked off the farm as a nurse, and Paul was the one who got to try his hand at tillage. Early on, peers and mentors told the Kaisers that tillage was central to farming. So that's what Paul did. Being constantly engulfed in dust was bad enough, but worse was the time he chopped up a snake. Nothing about tillage seemed right. The whole point of farming was to grow life, not kill things. On top of all the dust it generated, he discovered that after plowing the soil "turned to concrete and would lock up." Hardly any rainfall would soak into the ground, the opposite of what they needed given California's dry climate.

These early experiences led Paul and Elizabeth to wean their farm off the plow. They had also found a great young farmhand they couldn't afford to lose. He clicked with them and the direction the farm was taking. So that meant figuring out how to keep him employed over the winter. Farming all year also made practical sense. California offers year-round growing potential, and putting the farm to bed for the winter took a lot of work. Wine country also has hefty property taxes, and to pay their $20,000-a-year bill they needed more than two crops a year. Transitioning to no-till vegetables could keep money coming in year-round.

Their push to go no-till got a boost when a documentary film production came to the farm seeking shots of earthworms in vegetable beds. The first bed they dug into had worms but no photogenic vegetables. The second and third beds had the veggies but no worms. But the fourth bed had lots of worms and sexy vegetables. Paul realized that neither of the beds with worms had been rototilled, but both of the wormless beds had been. It was a striking revelation. Tillage and rototilling were killing the worms. Over time they switched from tractor tillage to rototiller to broadfork, and finally to no disturbance of the soil other than from transplanting vegetable starts into growing beds.

It took a while to figure out a system. Unlike larger industrialized farms that start the growing season with a plan for which field gets which crop and when, the Kaisers' approach is to decide a day or so beforehand. This way they are more in sync with the natural dynamics of the farm—pollinators, pests, and weather. Their way of doing things and the small scale of their farm allow them to "thrive on chaotic awareness."

Still, they do plan. It's just on the front end in their nursery greenhouse, sowing starts based on projected demand—what they think their customers will want over the course of a season. "It's all about where the crops will fit best at a given time. There are always about twenty factors involved: what was in a bed, what season, what crop is going in or coming out." And if they can companion crop and grow two or more crops in the same bed, they'll do that too.

They also over-sow the nursery because the worst sin is to have a field without plants in it. What do they do with extra transplants? Take them to the farmers' market.

In essence, Paul and Elizabeth are like a jam band of improvisational ecologists who ground their farming in observing nature. They improvise on the fly but know what they're doing. As Ray puts it, "It's not about tools—you have to understand the ecology and use the right tools."

By the end of the first year after they stopped plowing, the soil was already improving. And it just kept getting better. When they started farming, the organic matter content of their topsoil ranged from 1 to 3 percent in different locations around the farm. After going no-till, it rose an average of about 1 percent a year and then stabilized at between 10 and 12 percent, depending on which bed you measure. Paul and Elizabeth now manage the organic matter levels of their soil to keep it in this range. Just like the change that occurred in our garden, the increase in levels of soil organic matter at Singing Frogs Farm exceeds the proposed 0.4 percent a year goal for using agriculture to help offset global fossil fuel emissions on farmland worldwide.

The Kaisers didn't just stop tilling to increase their soil organic matter. As we walk among their beds, it's hard to find bare ground. Stout tomato plants rise from a thick layer of straw mulch in the middle of the first bed we stop at. They've cleverly used a viticultural technique—espaliering—to train tomatoes along a frame and make good use of vertical space in the narrow bed. A companion crop of lettuce is coming along on both shoulders of the bed. It's their second cropping of lettuce since March. They'll harvest $1,200 in lettuce off the 100-foot-long tomato bed twice before they harvest a single tomato. Similar pairings—a slower-maturing crop in the center of the bed sandwiched between fast growers—means they can harvest the quick growers (usually lettuce or other greens) every week of the year. This strategy allows the Kaisers to raise over 100 distinct varieties of 40 edible crops—among them, a dozen kinds of lettuce and 14 kinds of tomatoes, along with a half dozen broccoli and cauliflower varieties and numerous varieties of beets, cabbages, carrots, fennel, herbs, and flowers. Later-season crops like squashes will spill out of beds come fall.

Like the O'Haras, the Kaisers grow something in each bed all the time. This ensures that plants constantly capture solar energy and convert it into the exudates on which soil life thrives. In contrast to when they started, every bed now grows a minimum of three to ten crops per year. Paul says that in spring half the beds are companion cropped. The next bed we come to has fava beans planted in between Brussels sprouts. They find broccoli also does a lot better with a fava bean companion crop. Brassicas in general don't support mycorrhizal fungi, so Paul and Elizabeth plant legumes along with them. When we ask if they use cover crops, Paul reflexively replies that all their crops are cover crops. His point is a good one as the collective effect of all their crops protects the soil and feeds the biology that builds fertility.

It has been seven years since even a broadfork disturbed the bed we're standing on. We wonder what the soil is like and dig in with our hands. "Dig" implies effort, though. It's more like a gentle descent the way a spoon slowly sinks into a thick stew. The dark brown soil is still moist half an inch below the surface, even though the last rain fell in April, four months ago. Mulch and compost keep moisture in the ground, where crops and soil life can use it. After harvest, any remaining crop stubble gets cut at ground level and composted. The belowground roots and stems became food for soil life and eventually turn back into soil.

Everything on the farm is planted by hand. All but a few crops get started in the nursery. After a crop is harvested, any remaining parts get cut off at ground level and composted or left on the ground as mulch. Then a thin layer of compost is applied to the top of the bed, and the new crop is transplanted. Their method, much like the O'Haras', feeds soil biology and keeps nutrients cycling from mulch to soil and back to living crops. It also gives each crop a head start on weeds, making for minimal weeding. And again, just like at Tobacco Road Farm, weeds are not a major battle for the Kaisers.

They too use a lot of compost, making half themselves and buying the rest as municipal compost made from household organic matter. A common defense for not trialing such methods is that the compost costs too much. But Paul points out that they pay more money for seeds than for compost, and in any event, both combined add up to less than 10 percent of their costs—most of their outlays (about 80 percent) go for labor. They use horse manure, various kinds of vegetable matter, and sometimes sheep manure to make their compost, letting it mature for three to nine months. They also mulch with up to 200 hay bales each year.

Early in their transition to no-till, an organic farming consultant came to the farm and confidently opined that the levels of compost and mulch they were adding to their beds would send loads of nitrates down to their ponds and off the farm. In his experience, using the amount of compost the Kaisers did led to nitrogen-laden runoff. But there was a big difference between his experience on other farms and what the Kaisers practiced or, rather, no longer practiced—tillage.

Still, Paul and Elizabeth were concerned about the possibility of nitrogen runoff, so they tested the well and spring at the uphill part of their property and ponds throughout the property. The uppermost ponds had 6 to 8 parts per million (ppm) of nitrogen. They also tested the surface water outflow at the bottom of their farm four different times before and after heavy rains. It consistently tested out at 0 ppm: no nitrogen was leaking off the farm. It never got the chance. Not tilling meant more soil life that kept nitrogen cycling from one life form to another.

We head down to the bottom of the farm where the seasonal ponds are located, as Paul wants to tell us about something else that happened when they stopped plowing. For their initial three years of farming the ponds would fill with the first rains. Every autumn when they dried up, Paul would scoop four or five inches of eroded topsoil out of them and haul the wayward dirt uphill to spread back on the fields. But ever since they went no-till, it takes several storms for the ponds to fill with water and hardly any silt builds up in them. Now pretty much every drop of water that hits the farm sinks down into the soil, so there's no runoff and their soil stays on their fields.

Initially cucumber beetle outbreaks were a big problem at Singing Frogs. They still have them, but they have not lost a single plant to beetles in eight years. The crops are far less vulnerable, and "the disease pressure is now virtually nil." Paul and Elizabeth attribute these changes to improved soil health since they stopped tilling, as well as all the hedgerows and other noncrop plantings that provide habitat for pest predators, including those that gobble up cucumber beetles. While pests thrive in annual plants, predatory insects and birds need the year-round habitat that perennials provide. Not coincidentally, we notice loads of sharp-eyed birds hanging out in trees and shrubs around the borders of the vegetable beds.

Most farmers don't care much about their paths and roadways. But all over Singing Frogs Farm such areas, even between crop rows, provide erosion control and food or shelter for bees and other beneficial insects. Paul seeded them with over a dozen different kinds of plants, and about half a dozen performed well, mixing with native herbs and ground covers. We walk past pollinator-friendly oregano and flowers standing at the end of a crop row buzzing with insects. Such plantings bookend nearly every bed at Singing Frogs, reminding us of something missing from many conventional farms we've visited—the natural beauty of exuberant life.

Today seven people are working on the farm, and we all gather for lunch around a wooden table under the shade of a broad-leafed tree Paul planted when he and Elizabeth started out. Over a big salad with farm-grown cucumbers, zucchini, and squash, Elizabeth tells us about the commercial advantages of growing year-round. They get $4 a pound for early-season cucumbers but less than $2 a pound in the main season when other farmers also have them. Elizabeth says this strategy pays off economically and feels right for another reason: "It's important to feed your community year-round."

Although they don't use any chemicals, Singing Frogs is not certified as an organic farm. The Kaisers started the California paperwork three separate times, but the state wanted an entire input and output record for each bed for every crop. That would require at least a thousand separate entries for their two-and-a-half-acre farm. So they sell 90 percent of what they grow through a community-supported agriculture (CSA) group or at farmers' markets, with most of the rest going to restaurants. Their harvests all stay within a couple dozen miles of the farm. Most customers have toured the farm or heard the Kaisers speak at local events. They want to eat the kind of food Paul and Elizabeth are growing.

We continue touring the farm, and Paul steers us into one of the barns for a simple demonstration he has set up. He takes a soil sample from a conventional vineyard and holds it next to one from Singing Frogs. The vineyard soil is light tan, with a blocky texture that crumbles into powdery dust when lightly crushed by hand. The soil from Singing Frogs is dark brown with a sugary, yet cohesive texture perforated with roots and worm burrows. Paul drops a clod from each soil sample into separate jars of water, and we line up like schoolkids to watch what happens. The soil from Singing Frogs bobs a bit and ever so slowly begins to sink. It lands gently at the bottom of the jar as intact as it began. The vineyard soil disintegrates the minute it hits the water, falling as a curtain of cloudy sediment that collects in a layer at the bottom of the jar. Paul says their soil used to do the same thing.

Naturally we want to see some conventionally farmed soil for ourselves, so we visit a nearby vineyard and dig a hole. It has a whole different feel than the soil at Singing Frogs. Instead of a springy sponge underfoot, there is half a foot of powder-like dirt. Khaki clouds of dust explode around Paul as he moon-bounds in exaggerated strides between rows of vines. These differences have huge implications for water-holding capacity and soil erosion.

After easily pushing his soil auger through the surficial dust, Ray battles to drive it into the rock-hard earth below. But even when he throws his full weight into the effort, the auger bangs to a stop. Yet a gentle squeeze in one's hand turns a hunk of this hard dirt into fine, dry powder.

Ever the soil sampler, Ray is prepared. He places a coffee can with the bottom cut out on the ground surface and pours a bottle of water into it to measure how long it takes for half a liter of water to drain down into the dusty dirt. We watch, but it's worse than boring: the water just sits there. Three, four, five minutes and we only see the water level dropping a little. It takes another 20 minutes for all the water to soak in. What this shows is that most of the rain that lands on this field will run off instead of sinking into the ground. Ray also measures the soil temperature. Sitting above 100°F, it's higher than the 86°F air temperature and hot enough to kill soil life.

We return to Singing Frogs, where Ray places his soil auger on the surface of a bed and gives it a gentle turn. The auger virtually drills itself into the ground. It first hits an organic-rich layer called the O horizon. Mostly decomposing plant matter, it keeps the soil cool and feeds soil life. Ray keeps drilling and reaches the 2-foot mark with hardly any additional effort. When we dig our hands into the earth, it's moist below the surface and a little squish turns a fistful into a loose ball. When Ray measures the infiltration rate, it takes less than a minute for the same sized bottle of water to sink into the ground, more than 25 times faster than in the neighboring vineyard. This is equivalent to absorbing more than 20 inches an hour of rainfall. Because that much rain would never fall that fast here, all the rain that does fall on this field will sink into the ground, where it can sustain soil life and the crops. This is vitally important in drought-stricken California, the biggest producer of food crops in the United States.

The pioneering horticulturist Luther Burbank called Sonoma County one of the best places in the world to grow vegetables, and Singing Frogs proves him right. The Kaisers harvest well over $100,000 a year per acre of vegetables. Even good wine land only produces $10,000 to $20,000 an acre. And yet Paul says that in all of Sonoma County there are just 400 acres of vegetables among 60,000 acres of vineyards.

But how does the Kaisers' produce stack up? Customers are their best feedback, repeatedly buying the veggies Paul and Elizabeth grow. Some customers are skeptical at first. The produce from the farm comes with bits of soil clinging to carrots or a stray weed bunched in with some kale. But Paul and Elizabeth find that the same people usually come back for more, raving about the flavor.

There is a frequently tilled organic vegetable farm nearby, and we head over to take a look at the soil and decide on the details for crop comparisons with Singing Frogs. As we take in the scene, one thing is for sure: there is plenty of freshly plowed, bare dirt ready for planting. Like at the conventional vineyard, the soil on this organic farm is beat up, plowed into powder, and easily broken up with one's fingers. A thick, mulch-like layer of dust blankets the ground.

We settle on cabbage for a comparison crop and sample the soil in their main cabbage field. When we get the soil test results back, the organic matter level clocks in at 3 percent—the high end of what the Kaisers started out with. But there's also a new field recently converted to organic practices where the cabbages are not doing well. We decide to grab soil and cabbage samples from it as well. The soil organic matter comes in at just 2 percent in the test results.

When Paul looked up ten California organic farms he knows on Google Earth, the images showed that the soil looked much the same as on this farm—a lot of bare, plowed fields. No matter what type of agriculture, routine tillage leads to the same outcome—lousy soil structure and low levels of carbon and organic matter that ultimately undermine and harm soil life.

Do the regenerative methods of the O'Haras and Kaisers make a difference to soil health and the nutritional density of crops? The results of a particular type of soil analysis—the Haney test—reveal another part of the story. While organic matter is commonly measured in soil samples, it tells you only how much carbon is in the soil, not about microbial abundance and activity. This is where the Haney test comes in. It integrates aspects of soil biology and chemistry, taking into account soil carbon and nitrogen along with microbial respiration to gauge microbial activity—how fast the underground economy is spinning along. This analysis generates a numeric soil health score. The Kaisers' score is three times higher than the better of the two cabbage fields at the nearby tilled, larger organic farm, and the Kaisers have almost four times the soil organic matter. Likewise with the O'Haras' farm, which has a soil health score seven times higher than their conventional neighbor and more than three times the organic matter.

Crop quality tells the other part of the story. The Kaisers' cabbage had over 40 percent more vitamin K, a third to two-thirds more vitamin E, and 8 to 20 percent more vitamin C than the cabbage from the industrial organic farm. The Kaisers' cabbage also had over a third more calcium, about a fifth more potassium, and 5 percent more magnesium, along with less than half the sodium. But the biggest differences were in the phytochemicals and phytosterols.* Compared to the cabbage from the better of the fields of the large organic farm, the Kaisers' cabbage had a third more carotenoids, two-thirds more phytosterols, and 8 percent more phenolics. And compared to the cabbage from the poorer-quality field only recently converted to organic, the Kaisers' had more than twice the phenolics and phytosterols and almost 50 percent more carotenoids.

Relative to standard reference values in the USDA nutrient database, the Kaisers' cabbage had 50 percent more zinc and magnesium, and just a fifth of the sodium that one would expect to find in a commercially grown cabbage.† And while the USDA database does not include phenolics, a 2005 study published in the *Journal of the Science of Food and Agriculture* reported total phenolic contents for fruits and vegetables purchased from New York grocery stores, including cabbage, spinach, and carrots. The Kaisers' cabbage had almost two-and-a-half times as much as the grocery store cabbage. Spinach from both the Kaisers and O'Haras had about four times the total phenolics, and carrots from their farms had 60 to 70 percent more total phenolics. Soil health seems to matter, particularly to phytochemical levels.

---

* Phytosterols are plant fats that help lower LDL (low-density lipoprotein) cholesterol, the "bad" kind.

† USDA SR28, No. 11749, cabbage, common (Danish, domestic, and pointed types), freshly harvested, raw.

Soil Building

The common lesson from Tobacco Road and Singing Frogs Farms is that a sure way to restore fertility to degraded land is to build up and—crucially—maintain high levels of soil organic matter. That this would increase microbial abundance and activity is not surprising. Organic matter feeds soil bacteria and fungi, which increases not only microbial biomass in the soil but also plant biomass. When those plants are alive, they contribute carbon to the soil as exudates. And when dead, unharvested parts incorporated into mulches or compost return organic matter to the soil. This cycling of nutrients builds upon itself to make a self-reinforcing and incredibly productive system. But you have to let it work—and constantly observe the dynamics of life in your fields instead of panicking and reaching for poisons. You have to be willing to live with insect eggs and larvae of pests and predators for a bit, like on the O'Haras' kale.

The combination of organic and no-till practices that work so well for the Kaisers and O'Haras are not outliers. An impressive 21-year field trial started in 1992 just north of Munich, Germany, evaluated the effects of both organic versus conventional practices and full deep tillage versus minimal shallow tillage. Organic plots received organic cattle manure and either shallow or deep tillage. The conventional plots received nitrogen fertilizers and shallow or deep tillage. Minimum tillage in both plot types significantly increased microbial biomass. So did organic practices on tilled fields. Soil organic matter was significantly greater in the organic plots, as were bacterial and fungal abundance and activity. Bacteria particularly benefited from organic practices, whereas fungi benefited most from less tillage. But the biggest boosts came from the combination of organic fertilization and minimal tillage.

So if you want rich, diverse populations of both in your soil, feed the bacteria and don't disturb the fungi. Combining organic farming with minimal tillage to build soil organic matter can support a more diverse and abundant community of soil life. Benefits flow back to farmers through less chemical exposure and lower costs that translate into higher profits.

They also flow to the consumer in better food. We've all had crunchy but flavorless carrots, watery tomatoes, and other awfully bland vegetables. At the risk of sounding curmudgeonly, maybe they don't taste like they used to. If they did, we'd all probably eat more of them.

Researchers at Ohio State University ran an interesting experiment on potatoes in which subjects were asked to taste potato wedges to determine whether they were grown organically or conventionally. The same variety of Dark Red Norland potatoes was grown on neighboring fields in Wooster, Ohio, that had been organically or conventionally managed for at least several years. If the skin was left on, test subjects detected differences between the organic and conventional potatoes. They could not tell any difference between peeled ones.

So taste lay in the skin, which in the organic potatoes had significantly higher levels of potassium, magnesium, sulfur, phosphorus, and copper. The researchers also measured levels of solanidine, a phytochemical suspected to have cancer-fighting properties. Its concentration in the flesh of the organic potatoes was twice that of conventionally grown ones. Another study of potatoes found that their bitterness and astringency was related to their phenolic content and that vitamin C content had no effect on flavor. The takeaway? Phytochemicals and minerals strongly affect flavor.

While few studies have focused on how farming practices impact soil health, a 2020 study did just that for 20 farms in southern Sweden. The authors compared the soil on farms that used a range of tillage frequency, crop diversity, and organic amendments with adjacent unmanaged soils. They found that soil health on farms was generally poorer than in unmanaged soils and that among the farms less tillage, higher crop diversity, and more organic amendments translated into more soil organic matter, microbial activity, and aggregate stability. Across the board the study found soil-building practices improved soil health.

Regulatory models and incentives do not reward regenerative practices that build and safeguard soil health. Yet the Kaisers' and O'Haras' pioneering methods show that it is possible to rebuild soil health and enhance the commercial viability of small family farms. And though it might prove tempting to dismiss the potential for broader adoption of the methods they've developed, consider that instead of scaling up and extending their practices to large farms, we could scale out—to lots of small farms.

Paul and Elizabeth see the biggest roadblocks to small farmers as myths about conventional practices, lack of knowledge, and the high cost of land. While they think we'll always need big farms, they see small farms in rural areas close to big cities as the key to getting fresher, healthier food to city dwellers. Yet even though Singing Frogs Farm is a working family farm, Paul and Elizabeth don't get a tax break like many larger farms. They're too small.

More small and modest-sized farms focused on growing diversified vegetable and fruit crops in healthy, fertile soil would help address a long-running problem in contemporary America in providing a new business model for a rural livelihood. Compared to large-scale commodity crop farming, there are lower up-front and ongoing costs and the potential for far larger returns per acre. This same potential exists in urban areas and could help offset the higher price of land near or in cities where we need more farms focused on recycling composted food wastes and other urban organic matter. Both the Kaisers and O'Haras were pulling in enough income on small acreage to support a family and several workers.

Key to their success was that they grew plants in all beds at all times, monitored the organic matter levels of their soil, and disturbed their soil as little as possible in the regular course of planting or harvesting. While the specifics of their practices differed, both farms rested on a common foundation of building and maintaining soil health. And while the work is neither simple nor easy, growing food delivered rewards to each beyond the profits—restoring land and soil to a better condition.

But to grow a new crop of such farms we also need to change what we eat. Farmers plant what they can sell—and this is influenced by how the food industry makes and markets products we buy. All told, U.S. farms grow too few vegetables to supply every American with enough to meet dietary recommendations. More small-scale, soil-building vegetable farms in and around cities could help address the shortfall, especially if the produce tasted good and was nutrient dense—like what the Kaisers and O'Haras grow.

8 COMPOST CONNECTIONS

To be a successful farmer you must first know the nature of the soil.

—XENOPHON

Even though Eric Dillon hails from our soggy Seattle hometown, it's pretty clear his heart lies with Sandy Arrow Ranch. He invited us there one summer weekend to see the place and its soil. Tucked up against the eastern foothills of the Rocky Mountains in north-central Montana, dramatic ridges flank the wide-open valley through which Arrow Creek flows. While the far-as-you-can-see views enchant, the soil doesn't live up to the majesty of the terrain.

Since the end of the last Ice Age some 14,000 years ago, generations of buffalo and elk grazed the region, building up the fertility of the land. When Dillon heard the native soil in the area once had about 5 percent organic matter, he realized that modern grazing and cropping had run it down. In 2013, when he acquired the ranch, the soil had only about 1 percent organic matter—a fifth of what he'd heard the native Montanan prairie held.

When agronomic consultants advised him to regularly apply chemical fertilizers and pesticides to his fields, he questioned whether that was really the best way forward. The land had been productive for ages without added chemicals. Realizing his soil had drained batteries, Dillon started asking why we farm the way we do. He also began experimenting with compost, no-till, and cover crops to recharge his land.

As we toured the ranch exploring our mutual interest in soil health, Dillon showed us what he was up against. Digging into his fields, the dry, tan earth held no signs of life and little organic matter. Eric is a curious, quick study. Connecting his concerns about soil health at Sandy Arrow to broader problems of American agriculture, he wondered whether restoring life to his soil would grow more nutrient-dense food that could translate into better forage for livestock and better health for people.

We had the same question in mind, and Dillon was game to explore it. So we suggested a series of comparisons of crops and soil from neighboring regenerative and conventional farms across the United States. This could give us an idea of how much of a difference soil health could make to nutritional quality.

Across the Fencerow

The farms included Singing Frogs and Tobacco Road (though only soil samples were available for the latter) as well as eight pairs of neighboring conventional and regenerative farms.* Each paired farm had the same soil type, and all the farmers planted the same crop (peas, sorghum, corn, or soybeans) on an acre of their land in the same growing season. Crop samples were collected at harvest and sent to the Linus Pauling Institute at Oregon State University for analysis. Ray, the soil guy, collected soil samples from each farm pair at the same time during the growing season and sent them to a commercial soil lab for analysis.

* The paired farms are located in North Carolina, Pennsylvania, Ohio, Iowa, Tennessee, Kansas, North Dakota, and Montana. The regenerative farmers used a combination of no-till, cover crops, and diverse rotations.

Across the board, the soil on the regenerative farms consistently had more soil organic matter and higher soil health scores, with the Kaisers' and O'Haras' having the highest. Soils on the regenerative farms had 3 to 12 percent soil organic matter, whereas those on conventional farms had 2 to 5 percent (the Singing Frogs paired organic farm had 3 percent). Soil health scores for the regenerative farms ranged from 11 to 30, whereas those for the conventional farms ranged from 3 to 14, with Singing Frogs' paired organic farm at 9.* In terms of individual comparisons between each paired farm, the regenerative farms were consistently higher, with an average of twice the soil organic matter and three times higher soil health scores.

In terms of nutrient density, when averaged across all the paired farms, the crops from regenerative farms had a third more vitamin K, 15 percent more vitamin E, and 14 percent more vitamin B_1. The crops from the regenerative farms also had 15 to 22 percent more total phenolics, phytosterols, and carotenoids, as well as 11 percent more calcium, 16 percent more phosphorus, and 27 percent more copper. It appears that relative to conventional farming, regenerative practices yield crops with higher levels of at least some phytochemicals, vitamins, and minerals critical for our health. In other words, healthier soil with more soil organic matter and the life it sustains enhanced the nutrient density of crops.

Farmers using regenerative practices focus on building up soil organic matter in their soil. Composting is a favored method for smaller operations and highly variable. So how do you know if you've got the good stuff? It was time to head to Las Cruces, New Mexico, and visit David Johnson and Hui-Chun (pronounced Way-Chin) Su Johnson, a couple who took composting to a whole new level.

* A higher soil health score is better than a lower one. Soil health scores were determined by the laboratory analyzing the soil sample and reflect several factors, including microbial activity and levels of available nitrogen and carbon.

Compost Power

When we met up with them at a local coffee shop near the New Mexico State University campus, Hui-Chun's warm personality made us feel instantly welcome. David is skinny as a rail, and at first you might consider him to be on the quiet side. But once he got talking, we could see traits evident in careful scientists, an inner resolve and dogged tenacity. It peeked out from behind his bright blue eyes as he began telling us the story of how he and Hui-Chun ended up creating a composting system. They put their heads together to solve the problem of overwhelmingly bad odors David encountered while working on a research project.

Johnson's pathway into science was not the usual route. He'd spent 30 years working mostly as a general contractor with long stints in oil fields and logging. At the age of 50 he decided to go to college and in 2002 finished an undergraduate degree in biology. After taking another couple years to get a master's degree, he landed a job at his alma mater working with the Waste Energy Research Consortium.

He never thought he'd work on, let alone see value in, compost. The first half of that changed on his second day of work when his new boss said, "Have I got a project for you." It turned out to be a problem involving stinking piles of dairy manure. "Imagine my excitement," Johnson deadpanned. In New Mexico's arid climate, standard methods for producing compost often result in a salty final product. While the project was supposed to find good ways to use the manure, so far what it had shown was that the compost wasn't fit to put on anyone's fields.

Johnson kept working on ways to reduce the salt content. He went to conferences and took a composting class where he learned about moisture management and the importance of keeping compost oxygenated (aerobic) to cultivate the types of fungi that could clean up the salts. He found that making compost in long, linear piles (windrows) caused the surface to dry out while the center went anaerobic, suppressing beneficial fungi and concentrating the salt. He needed a different way to make compost.

As it turned out, more practical considerations provided inspiration. Due to his small project budget Johnson did all the windrow turning himself by hand. Moving all that compost was hard on his back, and it created a problem for Hui-Chun. Most evenings David arrived home from work smelling like manure and compost. There had to be a better way to keep compost aerobic and moist without tossing or turning it.

Experimenting a bit, he saw that compost became anaerobic about a foot down in the pile. So the challenge boiled down to how to design a composting system where the whole pile lay within a foot of air. He says if it hadn't been for Hui-Chun, he would have missed it. She suggested a simple solution: "Why don't we put pipes through it?"

As a contractor he'd done plenty of plumbing. This was doable. Together they designed a system where vertical perforated standpipes were used for aeration and as a conduit to add water to compost made of equal parts manure and shredded yard waste. It worked really well and produced high-quality compost nowhere near as salty as the windrow-composted manure. Little did he suspect that this fortunate turn would prove career altering.

He and Hui-Chun called their creation a "bioreactor." It consists of a cylindrical five-foot-tall structure made of a six-inch wire mesh frame covered with black landscaping cloth that sits on a standard shipping pallet. They started with a well-mixed ton of equal parts manure and yard waste loaded into the bioreactor. The next step was setting up a watering system on a timer and adding the all-important piping for aeration.

Their design placed lengths of four-inch-diameter perforated drainpipe vertically in the bioreactor, with one in the center and five circling around it halfway between the central pipe and the outer wall. The locations of the pipes ensure that an air source is within a foot of everywhere inside the bioreactor so the whole pile stays aerobic. The pipes themselves serve as forms for a day or two, after which fungi stabilize the material in the bioreactor and the perforated pipes readily pull right out with a gentle tug.

Their bioreactor design allows the pile to initially stay hot all the way to the bottom, reaching up to 160°F. Within a week it cools to under 120°F, and at 80°F worms (red wrigglers) get added. Irrigating the pile for a minute a day keeps the worms turning and mixing the compost, which in turn provides new surfaces for different microbes to decompose. After a year you get 700 pounds of incredibly rich compost from the original ton of material. The first time the couple tried their bioreactor out, it worked like a worm magnet. This impressed them both as the worms found the compost pile on their own, crawling up into it from the native soil, which they had thought contained very few worms. Best of all, the bioreactor didn't cover Johnson in manure or result in endless loads of stinky laundry.

You just set it up and let a diverse fungal community grow. If you turn it or disturb it, like in windrow composting, you keep resetting the fungal community. But if you leave it alone, like a good crockpot soup or stew left to simmer, the original mix of manure and yard waste gradually transforms into something utterly different—and much better than what you started with. Most importantly, the new compost system solved the salt problem too. The no-turn method provided habitat for fungal and other microbial communities to develop that could neutralize the excess salinity. The bioreactor had other advantages as well. The materials—shipping pallets and drainpipes—are inexpensive and easy to find. This makes for a versatile, modular design that can work on small farms in the developing world or be scaled up for commercial processing.

Having finally cracked the problem of what to do with dairy waste, Johnson took advantage of being a university employee to enroll in graduate classes for next to nothing. He picked molecular biology "kind of at random" and began taking courses toward a PhD. It proved a fateful choice.

At the same time, he still had one foot in the compost world. So he began to analyze the microbial community in the bioreactors and the compost they produced. Running a greenhouse test on plant growth, he compared the bioreactor compost to commercial compost brands and conventional fertilizers. While the bioreactor compost doubled growth, his data showed that conventionally measured plant-available levels of the big three elements thought to be most important for plant growth were not correlated with plant growth! How could this be? Johnson found that something else correlated with plant growth—the fungi-to-bacteria ratio.* More fungi translated into more growth. This was heresy and went against everything conventional agronomists taught. Who would believe that the amount and type of soil life was more important than chemical fertilizers?

So Johnson designed another greenhouse experiment with chili pepper plants to investigate the combined effects of the fungi-to-bacteria ratio and organic matter on plant growth. Using two types of "soil" as end members, desert sand and bioreactor compost, he mixed different proportions of them together to create five distinct growing media for chili peppers. Comparing the effects of the five mixes on plant growth, he found that the greatest growth was at the highest fungi-to-bacteria ratio. As this ratio increased, the plants produced more fruit. He also found that at low levels of soil organic matter more than 97 percent of the carbon fixed by the plants flowed down into the soil, building up the soil microbial community and increasing the overall organic matter content of the soil. But at higher levels of soil

* For anyone interested in statistics, the correlation coefficient (R^2) is a way to measure the strength of a correlation. A value of zero means no correlation at all, while a value of 1 means a perfect correlation. The greenhouse experiment R^2 for nitrogen fertilizer was less than 0.01, indicating no correlation, whereas the R^2 for the fungal-to-bacteria ratio was 0.88, indicating a very robust correlation.

organic matter and higher fungal populations, more of the energy the plant captured went into plant growth and fruit production.

It seemed to Johnson that the plants were trying to build up soil organic matter and microbial populations before investing in new growth. He thought the microbial community in the bioreactor compost gave the plants the boost they needed to reach that point. Notably, growth increased dramatically once soil organic matter got up to about 3 percent. A living soil was not only the key to healthy soil: it was more productive.

All in all, Johnson's findings from the chili pepper experiments pointed to a pattern. Plants invest in building up carbon belowground if they grow in soil poor in organic matter, but favor their own immediate growth if they germinate and grow in soil with higher amounts of organic matter. Think of it like a business. If you have to borrow funds to get started, you can't invest in expansion right away. Likewise, a plant that lands in poor soil initially invests in building up soil organic matter and soil life with exudates before directing energy to its own growth.

Following the greenhouse research, Johnson moved to a field trial thanks to a sympathetic entomology professor who offered him the opportunity to use a third of an acre on the campus farm. Johnson decided to compare growing a multispecies winter cover crop on a conventional control plot and a plot treated with a little more than a hundred pounds of bioreactor compost.

At the start of the experiment in 2009 the soil was pretty poor, with just 0.4 percent carbon.* In the first year, the light dusting of compost increased soil organic matter by more than half and produced more biomass—five times more compared to the plot without compost. For the next three years, he kept planting but didn't till or add more compost—and he kept measuring what happened to the soil.

* Soil carbon and soil organic matter are closely related, with carbon comprising about half the organic matter content. So in this case, the organic matter content was just under 1 percent.

By the end of the fourth year, soil carbon had tripled to 1.2 percent. It was still low but going up fast. The fungal community increased more than 20-fold, though from very low initial numbers. He noticed that the physical character of the soil changed too. At the start the soil was brick-like: a dropped shovel would ting off the ground. After a couple years the soil gently gave way underfoot. Each year as the soil improved, the cover crop biomass rose. The results were clear: re-establishing diverse biology and not disturbing the soil led to increased soil carbon and higher crop growth. By 2019 soil carbon was up to 1.7 percent, a fourfold increase.

The increase in plant-available mineral micronutrients in the soil was also striking. Copper and zinc increased by 40 percent and 62 percent, respectively. Molybdenum and iron increased more than tenfold. Magnesium and nitrogen almost doubled. Johnson had not added a single one of these minerals. But the bioreactor compost had radically increased fungal abundance and diversity. They were the force behind more plant-available micronutrients, unlocking them from soil particles and delivering them to the roots of plants.

Remarkably, all these changes began with a single application of bioreactor compost. It acted as an inoculant that worked like a master switch controlling a multitude of activities in the soil benefiting crops.

Curious about what the different microbes were doing, Johnson ran metagenomic analyses to identify which ones correlated best with plant growth. Out of thousands of species he found about 200 strikingly associated with growth.

While navigating these experiments Johnson continued his PhD in molecular biology and began tapping into a whole new set of analytic tools for measuring the microbial community composition as the compost matured. Dramatic, phyla-level changes occurred over the first three months, equivalent to the dominant species of an ecosystem shifting from clams to birds. But it didn't stop there. Things changed over the course of the entire year-long composting process. The number of species accounting for 80 percent of the total diversity quadrupled, increasing from almost two dozen after a month to about a hundred after a year. It was like building an ecosystem from the ground up. But in the microbial world where life spans are hours to minutes long, progress was far faster than in an ecosystem like a forest where trees mature over decades to centuries.

By the end of the composting process all the species were different from those at the start. A more diverse and complex community had replaced everything in the original compost. The bioreactor compost grew an incredibly diverse fungal community: it was a versatile inoculant full of possible partners for a crop to recruit. Here lay a key lesson. How compost is made makes a huge difference in the community composition of microbes in the soil.

Hui-Chun said that while most people want to rush the process, good compost takes time to mature. There is a gestation period. The process cultivates a diverse community of microbes that make an effective inoculant capable of creating conditions that support robust plant growth and build soil health.

How could so little compost work so well so fast? It wasn't about the sheer amount. Johnson didn't add anywhere near enough to explain the increase in soil carbon. That simply didn't pencil out. Something about the compost was stimulating enhanced plant growth and exudate production that soil microbes consumed and turned into organic matter.

Although the setting is very different, the rapid transformation in microbial composition of the soil is akin to how a single fecal transplant can quickly shift the whole population in the human gut.* The bioreactor compost appears to reinoculate the soil with a diversity of organisms that can persist over generations once re-established. To

---

* Fecal transplants are just what they sound like. Transplanting fecal matter from a healthy person into someone with recalcitrant *Clostridium difficile* infections has proven remarkably effective. We discuss the history and basis for this unconventional treatment in *The Hidden Half of Nature*.

keep a restored ecosystem thrumming, farmers need to follow inoculation with practices that support the new community—just as diet and lifestyle changes help the new community introduced through fecal transplants stick around.

Johnson still had the field plot at the university farm, so on our way out of town we stop by to have a look. We pass pecan orchards along the way with what look like dead zones in between the rows of trees. Concrete irrigation ditches feed water into orchards where the water pools instead of soaking into the dry soil. A tractor leveling the ground sends up a billowing dust cloud on the far end of a freshly plowed field.

We pull up to Johnson's field trial and watch a flight of butterflies dance across the plot. In the decade since initial inoculation he has grown cover crops and summer cash crops on the plot. At the end of the season they get mowed down, chopped up, and mixed back into the surface soil. It's a different version of what the Kaisers and O'Haras do but for the same end goal—improving and maintaining soil health through ensuring a food supply for the microbial herds beneath our feet.

The soil on Johnson's plot is dark and springy, whereas the khaki soil beneath the pecans growing a few feet away is hard, dry, and crusty despite the flood-style irrigation. Most of the soil on the research farm plots remains under half a percent carbon, a sea of dead dirt. How is this doing a service to farmers? Why aren't academic research farms trying to figure out how to restore soil fertility instead of how to grow crops in crappy soil?

Johnson sees reintroducing diverse communities of soil organisms as necessary to counteract an extinction event that modern farming brought to agricultural soils. Bioreactor compost reintroduces species, cover crops and crop-plant remains feed soil life, and the minimal disturbance of no-till conserves the restored habitat. No more junk food, no more poisons, just normal biology that feeds plants, soils, and microbes. For centuries, we've been living off a soil savings account built up over millennia. David and Hui-Chun have shown how we can reinvest in fertility and kick-start soil life using properly made compost that turns manure and other organic matter into medicine for the soil to get it back on its proverbial feet.

Growing Interest

Improving soil health is fundamentally about getting communities of soil life to build up their populations in ways that benefit rather than sap crop health. And one of the biggest drivers of this process is food. Compost like what the Johnson-Su bioreactor creates is ideal. It's like the prebiotic foods in the human diet—different types of fiber and other compounds in plants—that nourish human gut microbiota.

But what if your soil is low on soil life? Can you add to it? This is where probiotics—the organisms themselves—come in, typically in the form of inoculants. Chock-full of living microorganisms, well-made composts serve as both nourishment and inoculant for soil life. Once beneficial soil communities are re-established, you need to keep feeding and stop disturbing them, which means regular meals of compost and minimal tillage.

A number of studies in addition to Johnson's research have shown that compost can boost the abundance and activity of soil organisms that suppress crop-damaging pathogens. In particular, field trials have shown that organic matter–based fertility amendments like manure, biochar (charcoal), and of course compost increase soil microbial activity and that this reduces pathogen populations, although the particular effects depend on site-specific conditions and microbial communities. In addition, the degree to which compost and organic matter amendments work to suppress pathogens through boosting the nonpathogenic competition depends on both composting methods and the nature of the material composted.

Organic amendments also help crops increase mineral uptake. For example, green manure, composted animal manure, and biochar all tend to increase plant uptake of soil-bound zinc. In 2014, Swiss experiments on wheat found that using red clover and sunflower as green manure increased zinc concentrations in grain as much as adding zinc sulfate fertilizer. Organic matter not only enhances micronutrient uptake; it can reduce heavy metal uptake. Composted chicken manure, for example, reduces the uptake of cadmium in rice.

There are two general ways to promote the growth of mycorrhizal fungi: inoculate crops with them, or adopt farming practices that promote the growth of fungi already in the soil. The approaches are complementary. Inoculation can come directly from introducing microbes or indirectly from well-made compost.

It is well established that inoculation of crops with beneficial soil life can help kick-start biological processes at the root of soil health. Studies show inoculating fruit and vegetable plants with mycorrhizal fungi can increase both yield and quality—and influence the phytochemical content of crops. And the same goes for beneficial bacteria that associate with roots. Inoculating lettuce, spinach, carrots, and fenugreek with them not only induced higher total phenol and antioxidant content, but extracts from such plants better prevented oxidation in rat liver tissue. In other words, fungi and bacteria can influence plants to increase antioxidant levels in fruits and vegetables that in turn limit oxidative damage in mammalian tissue. This suggests eating plants grown in soil with beneficial microbes really is better for warm bodies, at least for rats, and most likely for us, as we'll delve into later.

Inoculation with certain mycorrhizal fungi can significantly increase the amount of different phytochemicals too, particularly phenolic compounds like flavonoids and other antioxidants found in artichokes, basil, garlic, peanuts, potatoes, lettuce, tomatoes, strawberries, barley, wheat, and corn. It stands to reason, therefore, that agronomic practices that promote mycorrhizal colonization could affect the phytochemical content of foods. And this opens the door to thinking about inoculating crops and soil to cultivate mycorrhizal fungi as a way to increase the nutrient density of food.

A 2006 UC Davis study reported that inoculation of tomato plants with mycorrhizal fungi increased zinc concentrations by almost a quarter to more than half in fruits and shoots, respectively. Likewise, a 2012 Italian study found that tomato plants inoculated with symbiotic fungi contained almost 30 percent more zinc and produced almost 20 percent more lycopene and antioxidants.

Other experiments using compost inoculated with a combination of mycorrhizal fungi and rhizobacteria significantly improved tomato quality, increasing the carotenoid content. A 2006 Indian experiment showed that inoculated tomatoes had more vitamin C and produced up to 25 percent greater yields under mild to severe drought conditions. Inoculation with beneficial fungi can significantly and quickly improve the mineral, vitamin, and phytochemical content of tomatoes. While this is one way to once again grow flavorful tomatoes, so too would practices that ensure such fungi thrive in the soil in the first place.

Onions are one of the most important dietary sources of flavonoids and other phenolics. Experiments comparing the growth and phytochemical content of different varieties of onions with various fungal inoculants from the genus *Trichoderma* found that the amount of fertilizer needed to obtain the maximum bulb mass was reduced by up to half for inoculated onions. At all fertilization rates, phenolic compounds were higher in inoculated bulbs, with the highest amounts coming from inoculated onions that received less than half to no fertilizer. In other words, fungal inoculation increased the phytochemical content and allowed for lower chemical fertilizer use. But the effects depended on both the variety of onion and the specific fungal species.

A diversity of fungi is even more beneficial. Here, we suspect, is why the Johnson-Su compost method works so well. It yields compost with lots of different kinds of fungi. A Czech study on onions found the best growth response and nutritional enhancement came from inoculation with both mycorrhizal fungi (the type that forms symbiotic relationships with plants) and saprotrophic fungi (organic matter decomposers). A mix of six different fungi almost doubled bulb weight, whereas inoculation with a single species did not enhance growth. Inoculation with a fungal mix also produced the greatest increase in antioxidant capacity and mineral content (magnesium, potassium, and sulfur) in the harvested onions. The synergistic benefit of both types of fungi makes sense insomuch as they produce complementary effects, with saprotrophic fungi enhancing cycling of organic matter and mycorrhizal fungi enhancing nutrient delivery to the plant, which builds more biomass, including roots. And more roots mean more exudates that in turn build more organic matter that feeds the saprotrophic fungi. Everybody wins.

Bacterial inoculation can also boost micronutrient uptake. Experiments at the Indian Agricultural Research Institute reported that inoculating wheat with rhizobacteria increased yields by more than 10 percent, protein content by almost twice that, and manganese by a third; it also more than doubled the iron and copper content of the grain and could increase zinc concentrations as well. Likewise, Saudi researchers found that rhizobacteria inoculation varied for different date palm varieties but generally increased soil organic matter, yields, and the phytochemical, vitamin, and mineral content of the harvested dates, with the copper content increasing more than tenfold for one variety.

That these increases would be enough to offset historical micronutrient declines suggests that disruption of soil life may well have played a role in those declines in the first place. They also found that inoculation increased the antioxidant and anticancer activities of the dates.

Combinations of fungal and bacterial inoculants appear to work even better than either alone. A 2017 Italian study of tomatoes found that inoculation with both plant growth–promoting bacteria and mycorrhizal fungi not only allowed farmers to greatly reduce chemical inputs but doubled plant biomass, increased vitamin C content, and improved quality and flavor. In a subsequent study, the same researchers found that a mix of mycorrhizal fungi and plant growth–promoting bacteria produced larger tomatoes and significantly enhanced the crop's phytochemical content. The doubly inoculated plants received almost a third less nitrogen fertilizer yet produced similar yields as the conventionally fertilized uninoculated tomatoes. They were also considered more flavorful and had higher levels of carotenoids than uninoculated ones. Similarly, a 2014 Indian study found that inoculating chickpeas with a mix of beneficial fungi and soil bacteria more than tripled total phenolic and flavonoid content over levels in untreated chickpeas. So it is not just bacteria or fungi that we want in our agricultural soils: it's both.

All this points to the potential for farming practices that promote mycorrhizal colonization and increase the abundance of beneficial bacteria to increase the mineral and phytochemical content of crops. Doing so would also grow healthier, more resilient, and more productive crops—and do it with less need for nitrogen fertilizers.

Commercially available microbial inoculants already exist, and research labs continue to seek marketable breakthroughs. And on a smaller scale, farms in tropical and temperate regions have devised a range of methods to produce inoculum. Just like the process for making sourdough bread, a farmer acquires a starter culture (such as from a sample of native soil) and introduces it to compost, letting it then mature in pots, drums, or piles before applying it to the fields.

While intensive composting methods are well suited for small-scale farms like the Kaisers' and O'Haras', scaling up to get organic matter back into the soil of large farms like Eric Dillon's ranch and the paired farms we sampled can occur in other ways too. The key to success is adapting organic matter delivery and cycling to the crops and regional climate of individual farms. Scaling up can occur through either a lot of small farms with access to organic matter from urban and suburban garden and food waste, or through large farms growing cover crops as green manure and reintegrating animal husbandry into crop production.

Regenerative practices that rebuild soil health, from the compost that the Johnson-Su bioreactor creates to practices like those of the Kaisers and O'Haras, serve another purpose too. They help ensure our bodies acquire adequate supplies of phytochemicals that underpin lifelong health.

OVERLOOKED GEMS

Food and medicine are . . . the front and back of one body.

—MASANOBU FUKUOKA

Tucked away in a quiet corner of the University of Washington, the physicians' garden doesn't get many visitors. Herbs and medicinal plants fill neat, geometric plots in this lovely spot of green sandwiched between academic buildings. On the occasions we've visited, it's rare to encounter another person. We've wondered whether campus doctors and medical students even know it's there, let alone are aware of the uses and effects of all those plants.

Yet doctors once had to know all about plants because medicines came from the botanical world and its phytochemicals. Long before we could visit a doctor's office and buy pharmaceuticals, healers and herbalists made it their business to know about the health-protective, curative, and healing properties of plants in the wild as well as those cultivated on farms and in gardens.

Science today is uncovering and explaining the health benefits of phytochemicals in the human diet. Not only do our crops possess inherent biochemistry to resist and deter pests and pathogens, but many phytochemicals work as preventive medicine when they land in our bodies. They play key roles in cellular cleanup, neutralizing toxins and curbing abnormal cell growth to fight cancer before it starts. And there is evidence that certain phytochemicals are connected to lowering blood pressure and cholesterol levels or prove helpful in preventing some chronic diseases, from arthritis and inflammatory bowel disease to Parkinson's.

Despite substantial research that points to important roles for phytochemicals in human health, they lack a solid place in the nutritional pantheon. There are no daily dietary recommendations for phytochemicals as there are for minerals and vitamins. Why, you may wonder, are phytochemicals not considered nutrients if they are so beneficial to our health? Nutrients are generally defined as the things essential for growth and survival. Carbs, proteins, and fats all fill the bill, as do vitamins and minerals. But what about the quality of that life? What about health? While assessment of the nutritional value of crops still generally neglects phytochemicals, what's coming to light is that in long overlooking their importance in plant and animal foods, we undermined our own health.

Green Goods

Yet we shouldn't rush to lump phytochemicals in with vitamins and minerals. They work through different mechanisms that support and protect human health. Recognition of their positive role in health continues to grow, and they are receiving serious medical interest.

Phytochemicals are often referred to as plant secondary metabolites. The "secondary" part of the term arose from initial beliefs that these compounds were waste products of plant metabolism. But by now we hope you realize that phytochemicals are actually quite important for plant health and defense.

The roughly 50,000 different phytochemicals identified so far represent only about a third of the number predicted to be discovered eventually. They contribute to the distinctive color, taste, and flavor of every fruit, vegetable, or grain you've eaten, as well as all the coffee, tea, and wine you've ever consumed. They also lend flavor to meat, dairy, and eggs. Tomatoes are well known for lycopene, which confers health benefits, especially when cooked. Anthocyanins give berries, cherries, plums, and grapes hues from dark red to purple. Terpenes give oranges their citrusy taste and are responsible for pine scents and the distinct aromas of cannabis cultivars. And while phytochemical levels vary greatly for any crop, how many of these overlooked gems you get in your diet reflects not only what you eat but how it was grown.

Phytochemicals fall into about half a dozen major groups. One of the best-studied groups—polyphenols—has over 8,000 different types. Some polyphenols can lower risk of coronary heart disease and help improve vascular health through influencing fat metabolism enough to inhibit formation of artery-clogging plaque deposits. Others retard tumor growth, reduce inflammation through antioxidant and anti-inflammatory effects, and help prevent certain types of cancer and dementia. Flavonoids, a subgroup of polyphenols abundant in soybeans, green teas, and dark chocolate, pack particularly potent antioxidant properties.

Many dietary phytochemicals come embedded in the fiber of plant foods. Since we lack many of the enzymes needed to fully digest plants, we have to rely on our gut microbiota to unlock the health benefits. Certain members of our microbiome readily modify fiber and phytochemicals into a range of compounds, many of which benefit our health through performing essential cellular maintenance, like serving on cleanup crews and anti-inflammation squads. Other bacterial metabolites affect our mood and cognition. In fact, the antioxidant effects of polyphenols in particular are due to gut microbiota that turn them into smaller molecules, making them more transportable in blood and easier for cells and tissues to take up. All this points to the importance of the human microbiome and the need to think more broadly about the nature of our bodies and the human diet.

Plants rich in flavonoids have been used in Eastern medicine for millennia, and Western medicine is now warming up to their antioxidant and anti-inflammatory effects as preventive medicine. A review of epidemiological studies found that greater intake of flavonols, a particular type of flavonoid, was associated with lower risk of heart disease and stroke. Flavonoids that inhibit oxidation are being explored for treating inflammatory chronic diseases, heart disease, and breast cancer. For example, isoflavonoids that bind with estrogen receptors of tumor cells can slow the progression of breast cancer. And a 2021 study that tracked about 75,000 men and women over several decades reported that eating a diet higher in flavonoids lowered risk of cognitive decline late in life.

Quercetin, a flavonoid pigment in tea, onions, grapes, cherries, berries, broccoli, cauliflower, and citrus, serves as a powerful antioxidant with anti-inflammatory effects. The human body cannot produce it. So what we eat provides our only supply. As quercetin inhibits the activity of enzymes associated with inflammation, it is widely used in the treatment of metabolic and inflammatory disorders. It also has antiobesity effects as it helps block uptake of glucose, inhibits fat accumulation in human fat cells, and spurs destruction of existing fat cells. Experimental studies have shown quercetin to reduce the risk of prostate cancer, inhibit cancer cell proliferation, and promote the death of tumor cells. Clinical trials also have found quercetin supplementation to reduce blood pressure, and blood levels of quercetin rise in proportion to the amount consumed in a person's diet.

Flavonoids also affect the color, taste, and health benefits of wines and are concentrated in the skin of grapes, with red wines containing more than whites. In particular, the anthocyanins that give red wines their color are thought to convey health-enhancing effects. Resveratrol is another polyphenol in red wine as well as blueberries. It is a powerful antioxidant and anti-inflammatory, and has anticancer effects that enhance the defenses of healthy cells and even promote the death of tumor cells. Moreover, a 2018 meta-analysis of randomized controlled trials of dietary supplementation with resveratrol found evidence of significantly reduced levels of inflammatory markers. There is growing evidence that resveratrol can help prevent or retard a range of health issues, including cancer, heart disease, diabetes, and inflammation-related chronic diseases.

Whether for resveratrol, quercetin, or other kinds of polyphenols, this all means you're likely to be healthier if you eat more of them. Will that guarantee health? No, not necessarily. But in concert with a well-fed microbiome these compounds help your body help itself.

While flavonoid levels vary greatly in crops, a 2000 review of their effects on mammalian cells found that diets rich in them can impart pharmacologically significant concentrations to body fluids and tissues. In particular, flavonoids play key roles in modulating the activity of inflammatory cells that serve on the frontlines of the human immune system—T cells, B cells, macrophages, and killer cells. Other kinds of polyphenols even have direct antimicrobial effects on gut pathogens.

Plants produce flavonoids as a defensive mechanism, with exposure to soil life or nutrient fluctuations, such as low soil nitrogen, increasing their levels. Herein lies a simple explanation for their higher content in organic crops. In contrast, conventionally grown crops that are spoon-fed nitrogen fertilizer and grown in soils short on life lack sufficient stimulatory cues to prompt robust flavonoid production.

Glyphosate use is another factor that influences the level of phenols in plants. Many are produced through the shikimate pathway—the one glyphosate shuts down. This does not bode well for the phytochemical content of crops grown on or near land where glyphosate is (or was) used to control weeds. Indeed, a 1992 review found that glyphosate inhibits the production of phenolics central to disease resistance in many plants.

Other phytochemicals also produce beneficial health effects. Carotenoids and phenols demonstrate anticancer effects in cell cultures and rodent studies. Higher carotenoid intake reduces the risk of bladder cancer in men. Phenols, terpenes, and glucosinolates inhibit tumor growth and lower the risk of cardiovascular disease and neurodegenerative diseases like Alzheimer's. Pilot studies and clinical trials suggest that a number of dietary polyphenols show promise not only for preventing but for slowing the progression of cancer. Phytochemicals can boost anti-inflammatory defenses and render malignant cells vulnerable to immune system attack. Anticancer effects have been reported for green tea polyphenols, resveratrol, curcumin, quercetin, genistein, and sulforaphane, among others.

Glucosinolates are a class of sulfur-containing phytochemicals derived from glucose and amino acids. They are abundant in vegetables of the mustard family like cabbage, Brussels sprouts, kale, broccoli, and cauliflower, yet vary greatly between varieties of each crop. Eating more glucosinolate-rich vegetables helps reduce colon cancer risk through activating detoxification enzymes that reduce DNA damage, inhibit cancer proliferation, and induce death of tumor cells (apoptosis). The phytochemical sulforaphane is another polyphenol found in brassicas and garlic that is known to inhibit DNA damage in colon cells. Studies show that a variety of phenolic compounds (flavonoids), sulfur-containing compounds (sulforaphane), and isoflavones (genistein) inhibit the growth of cultured cancer cells.

Animal studies show that phytochemicals tend to promote satiety and thereby reduce caloric intake. Some researchers think that polyphenols like resveratrol and quercetin mimic the beneficial effects of calorie restriction in extending the life span of mammals. One study even found that consumption of polyphenol-rich cocoa (think chocolate without the sugar) produced an aspirin-like effect on preventing blood clots. Perhaps we really should care about what goes into making our hot chocolate.

Despite established connections to health, there are few well-designed clinical trials studying the medicinal effects of dietary phytochemicals. Nonetheless, the evidence to date overwhelmingly points to beneficial effects on human biology. And how we farm offers a natural way to increase phytochemical levels in our food and thus the healthfulness of what ends up on our forks. The crop varieties we grow matter too, as phytochemical levels can vary greatly among those of any given crop. For example, plant breeding drastically reduced the amount of glucosinolates in common brassicas, lowering the amount of this cancer-fighting phytochemical in the modern diet.

Leaving phytochemicals in nutritional limbo masks an important fact. As we've seen, studies comparing organically and conventionally grown foods consistently find big differences in phytochemical levels. So how is it that phytochemicals confer so many health benefits?

Cellular Cleanup

To understand the intimate connections between phytochemicals and our health, we need to look at what goes on in the miniature biological metropolis of a human cell. When cells function normally, so do the tissues and organs they compose, which translates into health. Your cells have a daily routine. They need to extract energy from nutrients, excrete wastes they produce, make or repair things, communicate with friends, recognize and beat back foes, rest, and then wake up to do it all over again.

One way phytochemicals help keep our cells humming rather than stumbling through daily life is through influencing genes. Particular phytochemicals act like a shepherd, guiding special molecules into the cell's inner sanctum, the nucleus that houses DNA. One such molecule chaperoned into the nucleus is Nrf2 (cleverly pronounced "nerf-two"). It activates specific genes—and not just a few, but hundreds of them, many of which are directly involved in the day-to-day cellular housekeeping that enables our tissues and organs to keep working as they should.

Many of the genes that Nrf2 influences turn on antioxidant pathways, modulate inflammation levels, and spur immune cells that clean up and eat cellular waste and pathogens. Because of the interaction of Nrf2 with the human genome, it is considered a key player in longevity. That phytochemicals shape whether Nrf2 can get into the cell nucleus and directly influence a given gene helps explain why eating a diet rich in phytochemical diversity underpins good health.

But what should happen if phytochemicals aren't available when needed and Nrf2 remains uselessly parked outside the nucleus? Wastes pile up and get in the way of cells going about their daily tasks. This translates into trouble that spreads like fire. One cell begins to malfunction, nearby cells follow suit, and eventually the tissues and organs that they are part of suffer. This is how many diseases start, with cells that to some degree are unable to do the things they're supposed to do.

Much of the housekeeping that Nrf2 and similar-acting molecules set in motion is related to cleaning up the equivalent of biological exhaust. The trillions of cells that compose your body constantly burn energy to fuel their activities. This oxidation process yields compounds called "reactive oxygen species," or ROS.* Though produced by normal cellular activity, they can contribute to damaging tissues if they hang around inside your cells and body for too long. In such cases reactive oxygen species help lay the groundwork for chronic ailments, from cancer to cardiovascular disease and neurodegenerative and autoimmune disorders. For example, some types of ROS convert cholesterol to the oxidized form implicated in heart disease. In addition, they modify the chemical structure of proteins and enzymes, impairing how they function.

---

* Human cells produce several different types of ROS, among them nitric oxide and hydrogen peroxide. "Free radicals" is another term for ROS. Not all ROS are always harmful. Context and concentration are key, with some serving in various short-term cellular communication roles. Certain immune cells also produce ROS to kill bacteria.

But our bodies are clever. They employ an antidote for ROS—antioxidants. Many phytochemicals act as antioxidants, as do vitamins A, C, and E. So do zinc and selenium. In one way or another, antioxidants disable and dispose of ROS and thereby inhibit or prevent the oxidative damage at the root of many chronic diseases and other ailments.

Antioxidants do pretty amazing work, but cells get stressed when supplies run low. Excess ROS relative to antioxidant availability causes or exacerbates more than a hundred pathologic conditions, ranging from diabetes, to asthma and arthritis, to cancer.

So how do you ensure an ample supply of antioxidants? Take a drip-irrigation approach and eat them on a regular basis. Once-in-a-while binging or gorging isn't all that helpful. A steady supply is what you need. And on this point it's worth reiterating that the antioxidant effects of phytochemicals, especially polyphenols, depend a lot on fiber-loving gut microbes. For in the basement of our digestive tract—the little-loved colon—trillions of microbes do transformative work with whole plant foods in the human diet. More phytochemicals in our diet means our internal pharmacists can churn out more of the compounds that protect and help our bodies.

A 1987 study that tracked healthy European men in their 40s for seven years found lower mortality from heart disease and cancer for those with higher plasma levels of antioxidants. Several years prior, laboratory studies showed that vitamin C inhibits formation of nitrite-derived compounds linked to stomach cancer. Subsequent experiments using extracts of various fruits and berries found that higher levels of vitamin C and antioxidant phytochemicals (carotenoids and anthocyanins) inhibited growth of both colon and breast cancer cells. The researchers reported that vitamin C alone did not work as well as when combined with dietary phytochemicals.

Phytochemicals also communicate with cells, signaling that it is time to die should they start growing abnormally or cross into tumor territory. Sulfur-containing compounds in onions, garlic, and cruciferous vegetables have demonstrated such effects. This is not news. Documented effects of garlic extract on inhibiting the growth of malignant cells and tumors date back to the 1950s. Since then experimental and epidemiological studies have found that sulfur-containing compounds from plants inactivate carcinogens and suppress breast, skin, uterine, gastric, prostate, and colon tumors. These same phytochemicals also have anti-inflammatory and cholesterol-lowering effects in human subjects.

While farming practices influence levels of phytochemicals in crops, analyses of the nutritional content of foods for the most part tend to focus on macronutrients. And conventional produce generally looks pretty good through that lens. Recall that consistent differences between conventional and organic crops fall out most clearly in their phytochemical and antioxidant content—that is, in the levels of compounds that halt the start of or reverse ongoing disease processes but are not generally considered nutrients.

For all the reasons discussed above, from activating Nrf2 to their beneficial interactions with gut microbiota, phytochemicals are indeed overlooked gems. How we measure and think about the nutritional quality of food and the way it's grown has not caught up with expanding knowledge of what matters to health.

Eating for Health

In 1956 University of Nebraska medical professor Denham Harman proposed a theory of aging based on accumulating damage from the effects of free radicals in degrading healthy tissue and causing mutations. Invoking the idea of biological clocks, he thought that oxidative damage accumulating over time controlled the debilitating effects of aging. Colleagues initially scoffed at the idea, especially when early studies found certain antioxidants proved carcinogenic at high doses in laboratory animals. Many factors contribute to cancer and aging, but more recent studies demonstrated connections between free radical oxidants and cancer. Almost 40 years after Harman proposed his outrageous concept, he summarized evidence that vitamin C and phytochemicals acted as antioxidants with the power to counter and thereby limit damage from free-radical reactions. By then, many others who followed up on his initial idea found that free radicals did contribute to the onset and progression of cancer, heart disease, and Alzheimer's. Harman went on to surmise that reducing damage from free-radical reactions was the key to extending the human life span.

How might one do this? Eat foods with more dietary antioxidants, or eat less food. We are not suggesting that anyone "go on a diet," but rather to eat a diet of nutrient-dense foods that trigger satiety, as taking in fewer calories helps lower ROS levels. Substantial medical evidence supports the human health benefits of an antioxidant-rich diet, and dietary antioxidants have been shown to increase the life span of mice and nematodes. Likewise, eating fewer calories has been shown to extend the life span of rats, and not overeating defines a common thread among the world's "blue zone" diets from regions where people live the longest. So both approaches can work. In practice, however, Western populations do the opposite—eating lots of calories that contain fewer antioxidant and anti-inflammatory phytochemicals.

Greater dietary antioxidant consumption has been linked to lower formation of free radicals and to lower rates of many diseases, particularly various types of cancer and chronic diseases. While the therapeutic effectiveness of high levels of dietary antioxidants remains controversial for treating cancer, the case for preventive effects on a wide range of cancer types and chronic diseases appears overwhelming. But preventive effects of a plant-rich diet may reflect the combination of foods and phytochemicals in a diet as much, if not more, than the effects of individual ones.

Many phytochemicals now known for cancer-fighting or health-promoting properties—like sulforaphane in mustard family crops like broccoli and cabbage—were long considered antinutrients detrimental to health due to their bitter taste. Consumers didn't find the bitter taste of many phytochemicals appealing either. So the food industry routinely removed them through selective breeding or debittering processes. But as we'll see, the taste and flavor of bitter substances is a guidepost for healthy eating. Instead of focusing on figuring out how to prepare delicious, phytochemical-rich dishes, we've focused on making food bland, inadvertently undermining our bodily defenses.

Daidzein, one of the main phytochemicals found in soybeans, provides a striking reminder of the benefits of phytochemicals and how gut microbiota influence their effects. If you have the right microbiota dwelling in your colon, they'll transform daidzein into equol, a compound that at high-enough levels reduces the risk of prostate and breast cancer. This may be part of the reason why Japanese people who stick with their traditional, minimally processed, soy-rich diet have unusually low rates of these two types of cancer.

We suspect that science has a long way to go in fully understanding the complex communication that occurs in the human gut involving phytochemicals, the microbiome, and their beneficial effects on fundamental cellular mechanisms and pathways. If phytochemicals aren't essential for our bodies to function normally, why is our gut lined with dedicated receptors for them?

Humans came to rely on a variety of phytochemicals through our evolutionary relationship with the botanical world. For ages our ancestors foraged a variety of plant foods throughout the seasons, taking in a wide range of phytochemicals from different parts of plants. We sought out a variety of young green shoots and leaves in the spring and eagerly awaited summer's ripe fruit. In fall and winter we turned to starchy roots and tubers. The combinations of seasonally available foods also help explain the origin and ubiquity of multi-ingredient dishes like salads, soups, and stews across human cultures.

Our modern diet of course is quite different from our long-ago diet. The dearth—or abundance—of phytochemicals in today's diet depends on the state of the soil and how we fertilize crops and control pests. Every time we inadvertently hamstring the inner workings of our crops' green bodies, we shortchange our bodies as well.

Fortification programs are one way to make up for missing nutrients. They have helped ensure that people in more-developed countries consume adequate amounts of various vitamins and minerals. But thinking we'll fortify the human diet with phytochemicals isn't a simple proposition. This is the trap of single-nutrient thinking. The disease-prevention power of phytochemicals stems in part from synergistic effects that come from eating combinations of fresh, whole foods and gut microbiota that further enhance protective effects. Restoring soil health to jump-start the soil biology that spurs crops to produce a rich array of these botanical gems offers a promising way to support human health.

Edible Goods

Epidemiological studies generally find the lowest cancer rates among populations that eat a lot of fruits and vegetables on a daily basis. The anti-inflammatory and antioxidant phytochemicals in these foods are thought to underlie their beneficial effects on human health.

The potential effects of diet on cancer are well illustrated by a now-classic 1979 study from the chief of the epidemiology division of Japan's National Cancer Center. Among other things, the study compared rates of cancer for cigarette smokers who ate green and yellow vegetables (carrots, spinach, green peppers, pumpkin, leeks, and lettuce) on a daily basis against rates among smokers who did not. Habitual vegetable eaters had about half the mortality rate from lung cancer, an effect attributed to a diet with more antioxidative phytochemicals. This study established that diet affects susceptibility to cancer. Subsequent

European studies likewise found diets rich in a variety of fruits and vegetables reduced the risk of lung cancer among smokers and that consuming brassica vegetables in particular reduced colon cancer risk.

Other studies report that organic vegetables taste better, have a longer shelf life, and better suppress the cellular mutations that start cancer. What might explain what's so special about organic fruits and vegetables? As we saw earlier, they typically contain more antioxidants and other phytochemicals. Besides, as the studies referred to above come from countries where vitamin deficiency is not thought to be widespread, it points to a preventive effect of something other than vitamins—something like phytochemicals.

Over half, and by some estimates almost three-quarters, of all cancer is attributable to diet. And it is well established that obesity and greater consumption of processed meat promote inflammation and cancer proliferation. Likewise, alcohol consumption both causes inflammation and increases oxidative stress, producing reactive oxygen species that can prove carcinogenic if unchecked. Conversely, greater fruit and vegetable intake quells inflammation and reduces cancer proliferation. Put simply, increasing the vitamin, mineral, and phytochemical content of fruits and vegetables would benefit public health.

Inflammation predisposes a body to cancer and DNA, lipids, and proteins to oxidative damage that harm cellular functions. A 2014 paper in the *Journal of Nutritional Biochemistry* equated obesity to chronic, low-grade inflammation and reviewed how dietary polyphenols can prevent or partially reverse obesity-associated inflammation that contributes to human metabolic disorders. Foods with more anti-inflammatory and antioxidant phytochemicals can help reduce damage before it spirals out of control. How much more? It depends a lot on what you eat, how it was raised, and the soil it sprang from.

Epidemiological studies find that drinking green tea can reduce the risk of cancer. Once again, antioxidants are involved. Among beverages green teas are particularly rich in flavonols, a group of polyphenols that helps block formation of carcinogenic compounds and retard tumor growth. How does this work? Unlike black tea, green tea is not fermented. It is dried in a process that preserves its polyphenols, which can account for a third of its dried weight. Green tea flavonols in particular inhibit the activity of an enzyme that cancer cells need to successfully metastasize: without it they can't spread. Studies have even shown that inhibiting this same enzyme can reverse tumor growth in mice.

Tomatoes are another good example of a food with medicinal qualities. Consuming tomatoes reduces the risk of cardiovascular disease and certain types of cancer. They can contain high levels of antioxidants, such as vitamin C, flavonoids, and other phenolic compounds, especially lycopene, a powerfully antioxidant carotenoid. Lycopene influences key metabolic functions, including intercellular communication and the hormonal and immune system modulation involved in preventing many chronic diseases. Polyphenols in general and lycopene in particular have been found to both help reduce cancer cell proliferation and counteract the growth of estrogen-related cancer types (like breast cancer).

Does it really matter if your tomatoes have more flavonoids? Studies of more than 10,000 initially healthy Finnish men and women that tracked health outcomes from 1967 to 1994 found that a higher level of dietary flavonoids reduced the risk of chronic diseases. Specifically, higher flavonoid consumption reduced the incidence of and mortality from heart disease, the risk of asthma and type 2 diabetes, and the rates of lung and prostate cancer in men. Phytochemically rich tomatoes have been reported to help reduce cholesterol and blood pressure. Epidemiological studies find increased tomato consumption reduces the risk of cardiovascular disease and prostate cancer progression.

A feeding trial that had prostate cancer patients consume pasta with tomato sauce for three weeks prior to prostatectomy found that blood levels of lycopene doubled and prostate lycopene levels tripled compared to a control group that consumed their ordinary diet. Biopsies revealed that the tomato sauce–consuming group had lower levels of cancer progression and less oxidative DNA damage, as well as lower blood levels of prostate-specific antigen (PSA) due to, the researchers concluded, greater death of tumor cells. The surprised researchers suspected that lycopene acted synergistically with phenolic compounds (like quercetin) and other phytochemicals to produce the observed effect.

Soybeans also contain genistein, a polyphenol helpful in preventing a wide range of cancer types. It helps block signals that growing tumors send to get more oxygen and glucose, in effect starving cancer. This is thought to underlie why cancer rates increase dramatically among Asians who migrate to the United States and abandon a soy-rich diet. Other studies have shown that soy protein lowers blood cholesterol. These beneficial effects of dietary soy all flow from phytochemicals. As mentioned earlier, gut microbiota also influence the antioxidant effects of phytochemicals, especially polyphenols like genistein.

Many readers will no doubt rejoice that a review of studies found that high levels of chocolate consumption were associated with a roughly one-third reduction in cardiovascular disease and stroke. What do chocolate and fruits and vegetables have in common? Phytochemicals. Cocoa contains more phenolic antioxidants than most foods. And while chocolate is energy-dense from sugar, the benefits of moderate consumption of dark chocolate appear to outweigh the risks.

Overall, the available evidence indicates a protective effect of eating more fruits and vegetables. Yet the average American eats just over three servings a day of fruits and vegetables, well under the recommended five or more based on the associated reduction in the risk of cancer and degenerative diseases. On any given day in the United States almost half of the population does not eat any fruit and almost a quarter does not eat a single vegetable. What makes fruits and vegetables protective of our health? They are full of fiber and phytochemicals.

Not eating enough vitamins and phytochemicals is a global problem. A 2011 study of several thousand Brazilians age 40 or older found that fruit, legumes, and vegetables accounted for less than 3 percent of their food consumption. Virtually all of the subjects (more than 99 percent) had low intake of vitamin E, more than 90 percent were low in vitamin A, and 85 percent were low in vitamin C. More than half had insufficient zinc intake as well. The study also related that the country's typical diet would supply only a third of the antioxidants recommended for preventing chronic diseases.

In 2000, researchers fed 12 American women who had very high cholesterol levels a refined (processed) food diet for four weeks and then crossed them over to a phytochemical-rich diet for another month. The refined-food diet was representative of the typical American diet—a lot of conventionally raised, animal-based, and highly processed foods, with no more than two servings of fruits and vegetables a day. The phytochemical-rich diet centered on whole grains, nuts, and at least six daily servings of fruits and vegetables. Despite providing the same total caloric and fat intake on the two diets, the phytochemical-rich diet delivered less than half the saturated fat; half again more dietary fiber, vitamin E, and vitamin C; and five times more carotene. Just a month on the phytochemical-rich diet reduced cholesterol levels by about 15 percent and increased antioxidant activity by a third to two-thirds.

A subsequent 2005 study even found that diets high in plant sterols (a naturally occurring type of fat in plants), soy protein, and almonds work as well as statin drugs to lower cholesterol. Eating garlic helps too. A meta-analysis of clinical trials found that simply consuming half to one clove a day could reduce cholesterol levels by almost 10 percent. Diets rich in whole grains, legumes, nuts, seeds, and green, yellow, or orange fruits and vegetables contain antioxidant phytochemicals that reduce cholesterol. A diet rich in phytochemicals can also help protect against heart disease. A study that followed almost 50,000 men for eight years, from 1986 to 1994, compared health outcomes for those who ate either a diet characterized by high intake of fruits, vegetables, whole grains, and fish or a typical Western diet with high intake of processed meat, French fries, refined grains, and sweets. Those consuming the Western diet had a substantially higher risk of heart disease.

Phytochemicals derived from spices have been shown to suppress obesity-induced inflammatory responses, suggesting the potential for a more phytochemically rich and diverse diet to reduce chronic inflammation. Other studies show curcumin and green tea extracts help prevent age-associated disorders and cancer. Moreover, extracts from organically grown strawberries prevent the proliferation of cancer cells better than extracts from conventionally raised strawberries, an effect attributed to the greater concentration of antioxidant phytochemicals in the organic extract. And a study of over 70,000 Europeans older than 60 found almost 10 percent lower mortality for those who consumed a traditional Mediterranean diet with a lot of vegetables, legumes, fruits, and grains, as well as high amounts of fish and unsaturated fats (olive oil) and low intake of saturated fats (meat and dairy).

In overlooking the dietary importance of phytochemicals we've undermined our own health. We need to shift the way we think and consider how much better our bodies can fend off ailments and challenges with medical and defensive supplies always on hand. So the next time you reach for the perfect peach or strawberry to squish between your teeth, or savor the rich and complex flavor of a new wine, think about the character of the crops that bore this sumptuous food and drink. For they are a lot like people: it's what's on the inside that counts.

ANIMAL

SILENT FIELDS

Having land and not ruining it is the most beautiful art.

—ANDY WARHOL

Life came back below- and aboveground in our urban Seattle yard when we restored the soil to make a garden. Worms and beetles came first. Then a parade of pollinators returned to buzz among the flowers, vegetables, and shrubs.

All but two plants, that is. Purchased from a local nursery, each sported a couple dozen daisy-like flowers with a golden yellow center ringed with burgundy petals the color of old-style theater curtains. We never saw an insect land on them. When Anne inquired whether the plants might contain neonicotinoids, the nursery said the grower hadn't provided any information one way or the other.* We knew that neonicotinoid-infused plants are equal opportunity killers of both

* Neonicotinoids, often shortened to neonics, act as potent insect neurotoxins. When a seed is exposed to a neonic, the pesticide remains in the mature plant and gets incorporated into pollen, nectar, and other parts of the plant. Whether pollinator or pest, insects that later eat a part of the plant also take in the neonic, which can harm or kill nontarget species like bees.

insect pests and beneficial pollinators and predators. Had we unwittingly brought a deathtrap into our garden?

Neonics were supposed to reduce pesticide exposure for people, but this problem remains a serious issue for farmers and farmworkers. And although most of us think of farm settings as places pesticides are applied, about a quarter of pesticides purchased in the United States are used on gardens, lawns, and golf courses—enough that pesticide levels in urban streams are comparable to those of streams draining agricultural land.

Inhalation and skin exposure are not the only routes for pesticides to get into people. Residues can occur on or in crops or livestock that become part of the food supply. Despite the known health impacts of pesticide exposure, especially in children, global use of synthetic pesticides continues rising three-quarters of a century after their use exploded in the aftermath of the Second World War. Judging from their labels, acute exposure to pesticides is most dangerous. But what about chronic exposure people receive through consuming food or water containing traces of pesticides?

Every body processes toxins differently and also has different daily and cumulative exposures over a lifetime. While regulatory standards attempt to define "safe" levels for herbicides, pesticides, and nitrates that conventional farming introduces into our food, it is impossible to define a safe intake for all people, especially children. But we do know that exposure to fewer is better, and none is best of all.

Rising rates of birth defects and falling human fertility have been linked to pesticides. Across the United States birth defects fluctuate seasonally in concert with concentrations of both nitrates from chemical fertilizers and the herbicide atrazine. Pesticides have been found in human breast milk and are associated with adverse effects on male fertility and reproductive system development in animals and people.

And we know pesticides are getting into kids. Urine samples

from more than a hundred preschool children under five years old in the Seattle metropolitan area found that 99 percent had measurable amounts of at least one pesticide metabolite. Almost three-quarters of the kids tested positive for two or more. Children living in households where pesticides were used in the garden had significantly higher levels of pesticide-derived metabolites than those living in households with gardeners who did not use them.

A subsequent study of 23 children in Montessori and public elementary schools in Seattle suburbs found the pesticides malathion and chlorpyrifos in their body fluids and determined that exposure came primarily through diet. Conversion to a diet of organically raised foods dramatically and immediately decreased concentrations. Similarly, a study of preschool children in Seattle found those consuming a conventional diet had 6 to 9 times higher levels of organophosphate metabolites in their urine. Likewise, a study of over 100 Minnesota children ages 3 to 13 found pesticide residues or metabolites in the urine of all but 2 of those tested. And while exposure to pesticides varies greatly among children, those of farmworkers have significantly higher than average levels.

Pesticides are not just getting into kids. Testing of 1,000 adults living in the United States found pesticide residues in all but 20, with 6 pesticide metabolites occurring in more than half of those tested, indicating widespread exposure of the general public to pesticides or their residues.

Dietary intake is the most likely pathway for such widespread exposure among the nonfarm population. Little is known about human metabolism of the dozen most commonly identified pesticide residues. And people consume wildly different amounts of pesticides, as concentrations varied 100-fold between individuals. A study of more than 4,000 U.S. adults found substantially lower urinary organophosphate pesticide levels in frequent consumers of organic produce. And just as for kids, changing the diet of adults can rapidly alter their dietary intake of pesticides. An Australian study found a 90 percent decrease in urinary organophosphate levels upon switching adults from a conventional to an organic diet for a week.

Pesticide residues have been found in a wide variety of common foods and beverages, including cooked meals, drinking water, juices, wine, and animal feed. In Germany, for example, a thorough screening of foods found evidence for 361 different pesticide residues, with more than 60 percent of foods sampled having at least one. Some apples and pears had as many as a dozen. While none of us would buy a piece of fruit, bag of beans, or a loaf of bread with a skull-and-crossbones sticker on it, we're eating pesticides along with our food.

Certain foods tend to have more pesticide residues than others, with conventional crops generally having more than organic ones. A 2002 review of thousands of tests collected in the 1990s by the U.S. Department of Agriculture, the California Department of Pesticide Regulation, and the Consumers Union found that conventional fruits and vegetables were three times more likely to contain pesticide residues than organic ones. Almost three-quarters of the conventional samples contained at least one pesticide residue. For some crops more than 90 percent of conventional samples did. This comparison underestimates the difference between practices, as almost half the occurrences of pesticide residue on organic crops were legacy pesticides that root crops took up from the soil long after a field was last sprayed.

Although there is variability in people's exposure to pesticides and levels of pesticide residues on fruits and vegetables typically fall below regulatory values, the question of whether such levels adequately reflect health risks remains unclear. Some experts argue that regulatory limits greatly underestimate real health risks due to potential synergistic effects of multiple pesticides and chronic effects of low-level exposure over a lifetime, as well as from one generation to the next through epigenetics.

Just what might pesticides do in us? Of particular concern is that prenatal exposure has been associated with developmental problems in the first two years of life. Pesticides can pass through the blood-brain barrier, as well as the placenta to a developing fetus. A 2010 study of over a thousand children representative of the general U.S. population found that those with higher-than-average exposure to organophosphate pesticides had roughly twice the odds of attention-deficit/hyperactivity disorder (ADHD). In addition, childhood exposure increases the odds of other neurobehavioral problems, as well as leukemia and non-Hodgkin lymphoma (cancer of the white blood cells).

Researchers attribute the dramatic increase in non-Hodgkin lymphoma over the past several decades to higher pesticide exposure, especially to glyphosate. A 2019 meta-analysis of animal, lab, and epidemiological studies concluded there was compelling evidence for a link between glyphosate-based herbicides and cancer of the lymphatic system, a critical part of the immune system. Perhaps it is no coincidence that farmers face greater risk of this rare cancer, especially those in the central portions of the United States.

Moreover, pesticide residues have been associated with a wide range of gastrointestinal, neurological, respiratory, and reproductive effects and may contribute to the development of various types of cancer beyond non-Hodgkin lymphoma, as well as Parkinson's disease and endocrine-related disorders. A 2013 review concluded that abundant evidence supported considering typical pesticide exposure a risk factor for certain chronic diseases. A paper in the journal *Science* even reported that exposing female rats to high levels of fungicide resulted in three generations of male offspring with low sperm counts and thus reduced fertility. Along these lines it is worth noting that a controversial 2017 review found sperm counts among Western men had declined by more than half since the 1970s. Such concerns span the rural-urban divide given that agricultural land uses occur in watersheds supplying drinking water to major cities around the world.

No Buzz

If pesticides are as good as advertised at actually solving pest problems, why do farmers still need to use so much of so many? One reason is that habitual use of broad-spectrum biocides also kills pest predators, from flying insects and birds to soil-dwelling organisms. And with nothing to keep them in check, pests generally bounce back first. So the more pesticides you use, the more you need.

For millions of years before synthetic pesticides the botanical world deployed phytochemicals to fend off pests. Over the past century our growing reliance on pesticides unleashed a still unfolding insect apocalypse estimated to have killed off almost half the world's insects while contributing to a dramatic decline of grassland birds. Likewise, amphibian populations began declining rapidly in the 1950s.

Although their effects can be highly variable, pesticides indisputably affect soil life. A 2021 review of nearly 400 studies on the effects of pesticides on nontarget soil invertebrates found negative effects in 71 percent of cases examined. A key concern here is that pesticides reduce the abundance and diversity of beneficial mycorrhizal fungi. Soil fumigants like methyl bromide and formaldehyde are especially harmful and can eliminate fungi from a soil. This can affect what gets into food. Fungicide applications have been shown to reduce uptake of nitrogen, phosphorus, and zinc for various crops.

Pesticides are widespread in agricultural landscapes at levels that harm worms and other beneficial soil organisms too. A 2021 study, for example, measured levels of dozens of pesticides at 180 sites across farmland outside of Poitiers in west-central France. Samples were taken from conventionally and organically cropped fields, hedgerows, and grasslands. At least one pesticide was found in the soil at every site, and more than a third of the sites had more than 10. More than 90 percent of soils had at least one insecticide, one herbicide, and one fungicide. So did more than half the earthworms. A typical earthworm had more than 3 pesticides, and more than 9 out of 10 had at least one—one worm even tested positive for 11. More than 90 percent of the conventionally farmed fields presented a high risk to earthworms, leading the researchers to question the sustainability of conventional agriculture.

Herbicides and their breakdown products are common pollutants of surface waters. For example, a nationwide U.S. Geological Survey study found glyphosate and its primary, comparably toxic breakdown product—aminomethylphosphonic acid, or AMPA—in more than three-quarters of rain, soil, and river samples tested. A similar 2019 survey found glyphosate in more than two-thirds of rivers throughout France.

Glyphosate proves highly toxic to tadpoles, killing more than two out of three after one day of exposure and causing up to 100 percent mortality after three weeks of exposure. Experiments on more than a dozen species of larval amphibians found the potential for significant mortality at expectable levels of field exposure. Similar experiments found severely curtailed reproduction of tiny freshwater crustaceans resulted from chronic exposure to levels of glyphosate below permissible limits. Glyphosate also has been shown to be harmful to rat livers and to adversely affect the male reproductive system of rabbits and ducks. Studies have concluded that it can cause birth defects and neural damage to vertebrates in general and genetic damage to reptiles in particular. From insects to mammals it appears few animals can dodge the effects of the world's most popular herbicide.

What about us? Glyphosate occurs in food, though at low concentrations relative to its reported direct (acute) toxicity. USDA testing in 2011 found glyphosate residues in more than 90 percent of several hundred soybean samples. Did this trigger wider testing? Nope. The agency already considered glyphosate safe. Instead, regulators accommodated increased glyphosate use by increasing tolerance levels in corn, soybeans, and livestock feed. But as human studies have been virtually nonexistent, it appears that the definition of levels of glyphosate deemed "safe" in and on crops was set to accommodate conventional agricultural practices. Different levels are allowed in different crops for no clear reason, hardwiring the perception of no risk into regulatory practice because levels in food typically fall below allowable amounts.

Conventional agronomists generally regard glyphosate as safe to use due to low direct toxicity and a belief that it degrades rapidly in the soil. So until recently, hardly anyone, let alone the USDA or the Food and Drug Administration, tested for glyphosate or its primary decay product in food. Yet now that it has been found in food, blood, mothers' milk, and wine, researchers are starting to ask what it might do when it gets in us. A 2020 review found that more than half the species making up the core human gut microbiome may be susceptible to disruption by glyphosate.

It seems we've been running an uncontrolled experiment on ourselves. In 2016 a group of 14 scientists published a consensus statement of concern over regulatory levels of glyphosate tolerance being woefully out of date. They reviewed reports that found glyphosate was widely present in both water and food supplies, and summarized studies that suggested a basis for concern over human health effects at levels regulators considered safe.

Most press coverage—and lawsuits—on the human health effects of glyphosate focuses on cancer, and not without reason. Low concentrations of glyphosate have been shown to induce human breast cancer cell growth in laboratory studies through adverse effects on DNA repair, increasing mutation accumulation, and changing cellular metabolism. Moreover, laboratory experiments show that glyphosate and its co-formulants are toxic to human placental, kidney, and liver cell lines at concentrations lower than typically applied to agricultural fields. In other words, even low amounts in food may cause harm.

Another concern with glyphosate is what chronic low-dose exposure does to animal microbiomes. A German laboratory study documented secondary effects of glyphosate on the gut microbiome of chickens, finding that particularly virulent poultry pathogens (*Salmonella* and *Clostridium*) were highly resistant to glyphosate, whereas beneficial bacteria were harmed. The researchers concluded that feed containing glyphosate could perturb a chicken's gut microbiome enough to increase susceptibility to pathogens and disease.

Similar effects were noted in experiments investigating increases in rates of diseases in cattle associated with *Clostridium botulinum* (the bacterium that causes botulism in improperly canned foods). When laboratory experiments showed that glyphosate was toxic to lactic acid–producing bacteria that normally controlled *C. botulinum*, researchers concluded that ingestion of glyphosate-laden feed could undermine beneficial bacteria and fungi in the rumen and predispose cattle to disease. Subsequent experiments on the effects of glyphosate on the bovine microbiome confirmed that it inhibited beneficial microbiota and boosted rumen-dwelling pathogens. It seems indirect effects of glyphosate favor pathogenic bacteria associated with a disease that has become increasingly common among cattle.

If glyphosate scrambles the microbiomes of chickens and cows, what does it do in our gut? Recall that glyphosate is patented as both an antibiotic and a mineral chelator. So it's plausible that it could harm our microbiome and reduce or prevent uptake of minerals like manganese or zinc. Both bear on our health because antibiotics can change the composition of microbiomes and mineral deficiencies affect human enzyme function. These are among the reasons to challenge the wisdom of using glyphosate to "burn down" weeds and desiccate crops like wheat to make harvesting more convenient. The practice piles more glyphosate into the food supply.

While some studies have shown that pesticide applications can influence the mineral and phytochemical content of crops, the effects vary widely. For example, a Brazilian study showed that glyphosate use on glyphosate-resistant soybeans decreased crop levels of iron and a healthy fat you'll soon be hearing more about. And a compelling Norwegian study that tested batches of conventional glyphosate-tolerant and organic soybeans from Iowa found the organic soybeans had more protein and zinc and less saturated fat. In addition, the genetically modified soybeans contained high residues of glyphosate and its primary toxic breakdown product.

While the health effects and risk of chronic dietary exposure to pesticides remain controversial, there is no question that using less pesticide would reduce those effects. And improving soil health would help do that.

Roots of Resistance

Over the course of the early twentieth century, plant scientists came up with some jargon that is actually understandable—disease-suppressive soils. Experiment after experiment showed that plants grown in lifeless, sterilized soil to which pathogens were introduced succumb at rates far higher compared to growing in life-filled soils. Soil-building practices like cover crops and green manures are key to this dynamic as they support the kinds of soil life that displace, eat, or otherwise suppress pathogens. Studies have also shown that bacteria and fungi that promote plant growth stimulate root development and help plants weather droughts. Soil life can even make plants distasteful to insects or herbivores.

Plants also use certain phytochemicals, like glucosinolates, to control a wide range of soilborne pests, from fungi and bacteria to insects and herbivorous invertebrates. High phenolic concentrations in organically raised plants can make them less susceptible to pests and diseases. In one case, high leaf concentrations of phenolics reduced the size of female gypsy moth pupae, which then compromised the reproductive success of adult moths, decreasing their numbers. All in all, healthy soil helps plants resist pests and pathogens, reducing the need for pesticides. It's worth recalling here that glyphosate shuts down the shikimate pathway that plants use to produce phenolics. This may be a contributing factor in why a number of studies have found lower levels of phenolics in conventional crops.

Pesticides kill far more than a single target pest; they also kill many essential pollinators. This has prompted concerns over how to safeguard the pollinators essential for maintaining as much as 30 percent of global food production. In the United States beekeeping has declined by roughly half since 1950 due to insecticides, diseases, and loss of subsidies. A 2002 study in California found that conventional farms had greatly reduced populations of native bees, whereas organic farms near natural habitat support bee populations adequate to sustain crop pollination. The loss of pollinators costs conventional farmers a bundle, requiring large annual outlays for pollination services. Farming practices that are far less lethal to native bee populations offer a cost-effective way to keep pollinating crops.

Practices that enhance soil health extend to life aboveground as well. Reduced tillage provides stable microhabitats for spiders and other soil-dwelling predators. For example, a study from Zimbabwe found that tillage greatly reduced ground-dwelling, pest-eating spiders, while no-till farming and mulching did the opposite, leading to a larger and more diverse predator community. And a global analysis found that landscape simplification through adoption of monocultures and loss of hedgerows and trees from farms reduced pollinator and insect predator diversity, which in turn increased the need for pest control and reduced crop production.

So do we even really need pesticides? Organic farming generally leads to much higher abundance and diversity of birds and predatory insects than conventional farming. For example, a study of differences between conventional and organic production of the same variety of tomatoes on similar soils concluded that enhanced biological pest control could compensate for reduced use of synthetic fertilizers and pesticides. The study involved comparisons among 20 farms in the Central Valley of California where organic farmers controlled weeds through tillage while the conventional farmers used cultivation and herbicides. Most of the organic fields had been farmed organically for four to ten years using cover crops, worm castings, manure, and compost as fertilizer. The conventional fields received an equivalent amount of synthetic nitrogen fertilizer.

The study found that both microbial abundance and plant-available nitrogen were substantially higher on organic farms, as was the abundance and diversity of pest-eating insect predators. In the conventional fields, the incidence and severity of root diseases were consistently greater.

And although organic and conventional fields had similar total bacterial and fungal abundance, the community composition was strikingly different. Organic soils supported far more diverse populations of beneficial bacteria. While the overall abundance of herbivorous insects (pests) was comparable, the species richness and abundance of insect predators were more than 75 percent greater on the organic fields. In other words, unsprayed fields had more predators and thus about the same number of insect pests as conventionally sprayed fields. The message seems as clear as it is counterintuitive: routine use of pesticides increases pest abundance. Why? Pests rebound faster than what eats them.

Other studies further support the finding that reliance on chemical fertilizers increases the need for pesticides. For example, a greenhouse study of tomato plants found that those receiving higher amounts of nitrogen fertilizer had more nitrogen in their foliage. This in turn attracted beetles, which preferentially grazed on the more heavily fertilized tomato plants. Gorging on nitrogen, the beetles grew faster, got larger, and survived to adulthood in greater numbers. Another greenhouse study found that leaf-eating flies preferred to lay their eggs on tomato plants with high leaf nitrogen content. Doubling leaf nitrogen content tripled fly reproduction and doubled their survival time. And in yet another example, heavy nitrogen fertilizer use on wheat increased aphid reproduction and overall crop damage.

In other words, chemical fertilizers that increase the nitrogen content of crops translate into more, longer-lived pests. It appears that relying on soluble nitrogen fertilizers depresses plant defenses and produces bigger, healthier, more abundant herbivorous pests. Pests too are what their food ate.

Pumping up plant growth with nitrogen fertilizers created attractive meals for pests, cementing the legacy of growing dependency on continual pesticide use that unfolded over the twentieth century. This also explains the counterintuitive findings of a 2018 review that organic fields had lower reported incidences of crop diseases than conventional fields. The review found that composted plant debris and manure decreased the presence of yield-reducing pathogens that had become common under synthetic fertilizer use. Despite lack of synthetic pesticide use, organic management practices do not routinely trigger pest outbreaks. The takeaway is that pests unable to compete in soils with high microbial diversity can thrive in the low-diversity environment of conventionally managed fields. In short, conventional farming creates opportunities for pests.

A number of practices that build soil health help manage crop pests. The addition of beneficial fungi can enhance plant defenses against even aggressive pathogens. In northern Cameroon, for example, inoculation of maize and sorghum with certain mycorrhizal fungi reduced infestations of parasitic witchweed by roughly a third to half. Compost can help too. In West Virginia apple orchards, composted poultry manure provided effective weed control and boosted predator populations enough to greatly reduce tent caterpillar and aphid abundance.

Yet we shouldn't want to eliminate all pests. Minor insect damage and low-level exposure to pathogens increase phytochemical levels in crops. This likely contributes to why phytochemical levels in produce from organic farms tend to be higher than from conventional farms. Exposure to a little damage can tee up plant defenses—and infuse them with compounds beneficial to us. So we want a few insects to bother our crops. We just want predators to greet and eat most of them.

Curious about how feasible it is to wean conventional farmers off of pesticides, we headed off to visit an entomologist who left the USDA to start his own research and demonstration farm. He's showing that regenerative farming doesn't just reduce pesticide use. More insect predators mean more bucks for the farmer.

The Whistleblower's Farm

Flying in to Sioux Falls, South Dakota, in the lazy last days of summer, we gaze out the airplane window onto bare fields. The exposed soil reveals a pattern of dark-colored low spots between patches of lighter-colored ground. The dark areas look to be topsoil eroded off higher ground and piled into low depressions snaking across the landscape. The blotchy landscape looks sick.

What stands out even more, however, is the color of the ponds filling larger depressions left by ancient glaciers. Many glow green from fertilizer-induced blooms; others languish brown with suspended silt. Only a few sport a clean, deep-blue hue. The ponds tell the story of what went on in the fields around them as surely as the out-of-place topsoil.

We knew we'd arrived at Jonathan Lundgren's Blue Dasher Farm when we pulled off the freeway and saw a couple dozen sheep grazing in a small, electrified corral in front of a wood and stone house. Amid low, rolling hills the place stood out as an island of not-corn.

It didn't take long to notice all the life on his farm—chickens, pigs, sheep, bees, birds, and lots and lots of insects. And like Eve Balfour's farm, Blue Dasher is a scientific and economic experiment. Before he became a farmer, Lundgren worked as a research scientist. Now he's farming and doing research into science that university researchers and the corporations that fund them tend to shun—how to lay off agrochemicals.

But he also has to pay the bills. So he sells honey. A lot of honey. It's his best cash crop, his biggest revenue stream. Yet every year he loses all his bees. And so each spring he buys more.

A sea of conventional corn and soybeans offers slim pickings for bees and other pollinators. And what they do find gradually kills them with a double whammy. First is the declining quality of their food. Studies of goldenrod pollen indicate a substantial drop in protein content over recent decades that tracks rising atmospheric carbon dioxide levels. Second, bees that forage on pesticide-laden corn and soybeans transport poisons back to their hive trip by trip. As nutrition goes down and toxins creep up, a colony loses health and resilience.

So why did he buy a farm here of all places? Lundgren is on a quest for solutions to transform agriculture, and South Dakota sits on the frontlines where croplands, rangelands, and bees intersect. It helped that land was cheap. His 53 acres with a 5,000-square-foot home and outbuildings cost less than you'd pay for a backyard cottage in a West Coast city.

In khaki work pants, a bee-themed T-shirt, and a floppy-brimmed hat with vents, Lundgren leads us over to what used to be a milking parlor but now serves as his lab. One room, a customized facility for warming up cow poop, was truly unusual, perhaps unique. A whole wall stood stacked with shelves holding foot-tall sections of white, eight-inch-diameter plastic pipe full of cow pie. A light bulb hanging over each one heated the manure and drove dung beetles and other insects down through a funnel and into a bottle of ethanol. From here they could be identified and counted. So far, they've found over a hundred species of insects from South Dakota cow pies, far more than they find in conventional corn or soybean fields across the region.

Right after a cow deposits a fresh pie, flies scramble to breed in it. When the next generation of pesky, irritating flies hatches out, they readily find nearby cows. So in nature, grazers tend to move on. Rotational grazing emulates this behavior, frequently moving cows away from fresh pies. Before the cows return to graze again, insect predators and birds will have eaten the pests.

But not all insects hitched to cow pies are odious. Take dung beetles. They serve as nutrient distributors and recyclers in grassland and range settings, building soil health. Three different kinds dine on cow pies, albeit in different ways. Dwellers dive in and live there. Rollers grab some to take away. And tunnelers burrow down worm-like, mixing manure into the soil. All three were an integral part of a community of life that once followed bison herds.

Although herbivore dung comes out of the tail end of the digestive process, it's a beginning point for a great deal of other life. For dung is to beetles what pollen and nectar are to bees—a high-quality food source for maturing larvae and adults. But it turns out that pesticides are poisoning dung beetles as well as bees.

Avermectin is a biocide used to kill parasites in cattle. As it passes through the cow and into a pie, it also takes out dung beetles. Not surprisingly, Lundgren's crew is finding that more avermectin in cow manure means fewer dung beetles to recycle manure into soil organic matter. So soil health takes a hit when the activity of dung beetle dwellers, rollers, and tunnelers declines.

Lundgren never thought he'd become a farmer himself. Perhaps that's part of why he isn't mired in conventional thinking. Raised in suburban Minneapolis, he worked at the Minneapolis zoo in middle school and high school as a zookeeper's aid and focused on domesticated animals. In his freshman year at the University of Minnesota, Lundgren took an entomology class and was bugstruck. A quarter century later these formative experiences with animals came in handy when he began farming.

He moved to Duluth to finish college and worked as a gypsy moth tracker before getting his master's degree in biological pest control.

Lundgren says he "hated school, but was good at it." Next he moved on to a doctoral program in entomology at the University of Illinois and began working on risk assessments for corn genetically modified to incorporate an insect-killing gene found in a common bacterium (*Bacillus thuringiensis*). The killer corn, known as Bt corn, soon taught him far more than he wanted to know about industry involvement in science.

At first everything he measured indicated Bt corn was safe for lady beetles. The standard timeline for the toxicology assays that Lundgren followed was ten days, and after monitoring the lady beetles for that long all looked good. Everyone was happy with the result—Lundgren, his adviser, and the company funding their research. But things changed when he ran the same assays on a different proprietary Bt corn product for an extra, eleventh day. When he checked on the lady beetles, they were all dead. So, Bt corn was lethal to lady beetles after all. Despite this important finding, Lundgren learned that he couldn't publish the results of his research. Unbeknownst to him, his adviser had signed a nondisclosure agreement in order to secure funding. He calls the experience an education in "how the game is played." If nobody knows, there's no problem.

After completing his thesis he landed a job at the USDA research lab in Brookings, South Dakota. Glyphosate-resistant weeds were becoming a problem, and he wanted to research how to use cover crops instead of herbicides to manage weeds. So he started looking into intercropping or cover cropping to suppress weeds as well as soybean aphids, a major insect pest. He visited farmers who were making the practices work, but he couldn't replicate what they were doing. Worse, he demonstrated that he really didn't know what he was doing as a farmer. His soybean crop failed.

So he began to work on natural enemies of corn rootworm, another major pest. Diving back into entomology led him down the road to regenerative agriculture. He started looking into the stomachs of insect predators and found loads of rootworm DNA. The only place it could have come from was loads of rootworms, so the more insect predators of rootworm in a field, the less damage to the corn. To Lundgren the conclusion was clear: keeping predators on the job provided effective pest control.

Then he went to a no-till farming conference where the farmers hardly mentioned corn rootworms. Why? They had none to speak of. Insect pests were not on their radar. Instead, these no-tillers were talking about soil health, their favorite cover crop mixes, and crop rotation schedules. Rebuilding the fertility of their soil allowed them to not only lay off the pesticides but greatly reduce their fertilizer bills.

Intrigued, Lundgren convinced one of the farmers to let him inoculate a field with a thousand rootworm eggs per linear foot to mimic an infestation. None of them hatched out. That no pests emerged in a pesticide-free field flew in the face of conventional thinking. The place should have been crawling with them. When Lundgren laboriously counted insects from the inoculated field, he estimated the predators of corn rootworm and other soil pests at about a billion predators per acre. All those rootworms he introduced never stood a chance. This was the way the eat-and-be-eaten nature of soil worked—if you let it.

While he never did write up that informal experiment, Lundgren moved on to better controlled and replicated studies. He realized that most researchers focused on a single species. Pest people focused on pests and biocontrol folks focused on predators. No one looked at the interrelatedness of insect communities—who ate what, when, and where.

Studying predator-prey relationships in agricultural fields looked like the way to make real progress. So he started doing comprehensive inventories of insect communities and found that as diversity goes up, pest numbers go down. But it wasn't just reducing pesticide use that mattered to insect diversity; there was an indirect route too. Cover crops provided habitat for predators, and when they had places to live, they preyed on pests. Food, shelter, and places to raise young, that's the good life for a predator seeking things to eat. Lundgren came to the conclusion that farmers were creating their own pest problems when they nuked predators with pesticides and grew the same crops all the time. Because prey populations rebound before predators, conventional farms dished up an unguarded self-serve buffet for pests.

In 2011 his research was going well, and the USDA honored him as one of its top scientists at a White House ceremony. Around this time Lundgren moved to working on bees. He did risk assessments on neonicotinoid pesticides, documenting the damage they caused to non-target insects (predators). He also showed that insecticide-treated seeds didn't increase yields and failed to reduce soybean aphid abundance. This was all very inconvenient, and certain folks were not happy with him. He was pissing off some powerful companies.

By 2014 things started to go downhill for Lundgren. He says that when he was scheduled to speak at an Environmental Protection Agency scientific advisory panel on pesticide risk assessment, a representative for a company that makes pesticides approached him with an offer to buy a lot of lady beetle eggs for a princely sum of $700 apiece for a study they wanted to conduct. Lundgren considered a dangled promise to buy even more eggs the following year a thinly veiled bribe to follow the company line. Instead, when his time to speak came, he boldly proclaimed the company was misrepresenting and underestimating the risk of their product. It made no difference. The Environmental Protection Agency ignored his assessment.

Then things got ugly. After giving an invited talk at the National Academy of Sciences about risks associated with genetically modified crops, a supervisor flagged one of his travel documents as not being signed. Lundgren says he simply forgot, but he was suspended for two weeks and told to stay in line and stop talking to the press.

He'd never thought about becoming a farmer himself but knew he couldn't keep working within the conventional research system: he asked too many questions and talked too freely about his findings. If he stayed where he was, he couldn't do the studies he thought needed to be done because the funding and infrastructure in agricultural science mostly supports conventional interests. The nonconventional farmers he'd met were innovative and exciting. Why not start a research farm and do science related to regenerative agriculture?

For the next couple of years, he says he faced daily harassment from supervisors. Higher echelons of the bureaucracy apparently did not appreciate his insights—or his proclivity to produce data that backed them up. He didn't stop. His crew showed that the nearly ubiquitous practice of growing sunflowers from neonicotinoid-treated seeds neither reduced pest levels nor boosted crop yields. They also showed that general use of neonics contaminated strips of wildflowers planted to support pollinators enough to undermine the health of worker bees. Hearing that flowers could be deadly reminded us of that pair of odd, insect-free plants in our garden.

At the same time Lundgren's colleagues at universities related having experiences like he'd had during his PhD work. Concluding that this happened everywhere and the system was broken, he pushed all of his chips onto the table and quit.

Would farmers and beekeepers support an independent research farm? He began crowdsourcing funding for Blue Dasher Farm. No one had tried this before. But it worked. One donation at a time, he landed several hundred thousand dollars, ending up with a better start-up package than he'd get at most universities.

When a South Dakota State graduate student needed a thesis project, Lundgren suggested comparing soil organic matter, insect pest populations, crop yields, and profits for cornfields on regenerative and conventional farms. On ten pairs of farms spanning Nebraska, North Dakota, South Dakota, and Minnesota, she found corn pest populations were ten times higher on conventionally insecticide-treated farms than on insecticide-free regenerative farms. It was another heretical finding: relying on insecticides to manage crop pests ensured ongoing pest problems. They don't put that on the label.

The comparison also found that profitability was unrelated to grain yield and yet correlated with soil organic matter. While it cost a lot more to get the highest yields due to high monetary outlays for fertilizer and pesticides, organic matter–rich soil reduced such expenses on a farmer's balance sheet. Healthy soils made for healthier profits.

In other studies Lundgren found crop rotations could reduce soybean aphid populations to just a quarter of prior levels. He even proposed that adding high-value, highly profitable oil seeds to typical corn-soybean crop rotations would help reverse bee declines by providing enough nectar to support healthy colonies.

Lundgren also told us of an ongoing comparison of regenerative and conventional almond groves in California. The soil in the regenerative groves had almost a third more organic matter and six times the capacity to absorb water than the soils of the conventional groves. And while both conventional and regenerative groves had the same amount of insect pest damage, the regenerative groves had almost a third higher insect diversity, six times the insect biomass, and far more predators. Lundgren related how when one of the conventional pesticide-using farmers saw the initial data on pest abundance, he simply didn't believe the numbers, responding incredulously, "You mean I'm spending all that money for nothing?" Relating this, Lundgren laughed, saying it was going to take more than data to change things. You have to change minds.

Lundgren takes us to a section of his farm that at first glance looks like an overgrown field. We stand for a minute, and he points out the assortment of flowering prairie perennials between young fruit trees—a camouflaged orchard-to-be of apples, pears, peaches, and plums, with raspberries, strawberries, and grapes in the understory. Any diseased fruit his young trees cast off gets eaten by chickens and pigs rooting through it all. As in nature, nothing goes to waste.

And nature responds. Blue Dasher Farm supports loads of wildlife—badgers, deer, coyotes, skunks, and bird diversity like we'd never seen on a farm. His fields are teeming with life, unlike the conventional fields that surround them. But for his bees that forage up to miles away, the surrounding landscape presents a chemical minefield.

Like the bees, his conventional neighbors aren't doing well. They've been losing money since corn prices peaked in 2012. With no way to make ends meet growing conventional corn, they sell at a loss and take their crop insurance payment. Still, everything for miles around remains planted mostly in corn and soybeans. Lundgren thinks it's time for something different.

Although the Dakotas are the top honey-producing states in the country, beekeepers there, as elsewhere, are going out of business. Lundgren says saving the bees will take weaning farmers off pesticides: "More habitat won't work if we keep routinely spraying everything." He suspects that unregulated ingredients in commercial formulations factor into why glyphosate-based herbicides harm bees. Because regulators consider such ingredients safe until shown to be toxic, their effects in a mixture remain unknown. No one looks. Yet insects are exposed to full formulations, not just the regulated active ingredients.

So are people, and it seems to matter. In 2013 French scientists studied the effects on human cell lines of glyphosate-based herbicides that included various efficacy-enhancing compounds. Such adjuvants are generally considered inert, but the researchers found that all nine formulations tested were more toxic than glyphosate alone, damaging mitochondrial activity, cellular respiration, and membrane integrity at concentrations typical of environmental and occupational exposure.

While glyphosate and other pesticides are very effective at doing their immediate job, maybe the job itself is shortsighted, even futile. If we're managing the pests, we are managing the wrong species. We need to better manage ourselves.

So how do we get off pesticides? Lundgren sees crop diversity and building healthy, life-filled soil as the key. He notes that the same health problems taking out bees are increasingly affecting us—autoimmune diseases, learning disabilities, and food intolerances. Using less is a logical way to lower pesticide exposure for bees and us.

Forgoing tillage and herbicides leaves Lundgren with burning and grazing as the tools he can work with to restore his prairie. So he uses both. First thing in spring he burns a field. His sheep graze what comes up, and then he plants right into it. By grazing his sheep in the burned field, he makes money off his weeds instead of spending money on herbicides. Sheep and bees both thrive on the native grasses of his agricultural prairie. So he sells lambs and honey. Every square inch of his farm produces more than one revenue stream.

Lundgren corrals his sheep within a mobile 50-by-50-foot electric fence. It takes two days for a couple dozen sheep to graze off three-quarters of the fenced-in prairie biomass. Then he moves them and the fencing on to another patch of lush vegetation. He sees his prairie as a never-ending sheep buffet. So do his sheep.

Blue Dasher Farm is also a great place to be a chicken. Lundgren keeps his laying hens until they die naturally. They usually live about three years, and the old ones teach the younger ones how to be a real chicken. They lay eggs all over the farm. He collects them from known spots and doesn't always find them all. Sometimes a hen leads a group of hatchlings out of a thicket. Then he's got more chickens!

Even though he designed the farm to be a bee paradise with loads of pollinator-friendly habitat, he loses all his bees every year. No matter where he places the hives, they're too close to the neighbor's conventional fields. "All my neighbor has to do is put one fungicide on each year and all my bees are dead. I just need one strain that can survive that chemical bottleneck, but I haven't found it yet." His queens used to live for three years. Now they last six months.

A steady breeze makes the day seem cooler than it is as we walk over to the neighboring conventional corn and soybean fields. The soybeans are knee high, the corn up over our heads. There are hardly any dragonflies and just a solitary butterfly flitting around. Crickets are the only insects on the ground, in stark contrast to Lundgren's ground crawling with legions of dark beetles. Though insects thrive on his farm, he has no problems with the mosquitoes that plague his neighbors. His secret? A squadron of blue dasher dragonflies feasts on any bloodsuckers straying into his airspace.

Lundgren bought the farm from a rich, casual landowner who threw in a couple dozen neighboring acres of native prairie and wetland. They had no economic value to him. Lundgren considers these worthless acres the crown jewels of his farm.

When he first got the place, Lundgren thought there were too many invasive species in his prairie. So he wanted to fix it. State agencies told him he could spray it with herbicide but sternly advised him not to burn or graze it. He ignored them and did both anyway, perplexed by why they'd let him spray it with poisons but object to doing what nature had done for millennia.

The spring burn woke up native plants. Now the native bluestem grasses are coming back. After three years his farm hosts a palette of flowers not seen in the area for years, as well as porcupine grass, milkweed, prairie rose, switchgrass, and goldenrod. As we stand and talk among the swaying grasses and a striking diversity of flowers, Lundgren bends down and parts the tall grass to show us something extra special—a small patch of eye-catching, dark-purple flowers pushing upward. It's downy gentian, an indicator species of native prairie plant communities. It was only after he started spring burning that Lundgren began to see these luminous flowers. Now they're moving in, and the weedy brome grass is almost gone. The prairie is coming back to life while Lundgren feeds his livestock and bees.

How much pesticide can regenerative farmers eliminate? Lundgren is sure they could eliminate most if not all of it. Farmers convinced they need to spray are just doing what experts with experience in conventional systems advise. But he thinks that if farmers adopt practices that transform their system, they can greatly reduce their need for insecticides, herbicides, and fungicides. Lundgren advises interested farmers to start with a corner of their farm and only expand it out once they figure out how to make regenerative practices work on their land. After just a year, he expects, most farmers will start to see benefits to the soil that keep growing year after year.

While most American farmers would not recognize Lundgren's farm as a farm, it's both profitable and buzzing with life: dancing dragonflies captivate our eyes and jungle sounds fill our ears standing at the edge of his agricultural prairie. How, he wonders, did we get to the point where modern farms seem devoid of life? As if in reply, a splendid orange and black butterfly wings aloft on the breeze.

When Lundgren worked out typical per-acre costs in South Dakota for Bt corn seed, herbicide, and fungicide, his estimated total bill came to $167 per acre. For a 400-acre farm this translates into about $67,000 a year on pesticides, $10,000 more than the state's median household income. He sees South Dakota farmers throwing away the equivalent of a second income that could bring their kids back to the farm.

Instead, he says, they're sending all that money to large corporations via their pesticide dealer. Only most of them are sending a lot more than that. The average size of a farm in South Dakota is more like 1,400 acres, which at $167 an acre pencils out to nearly a quarter of a million dollars a year for the average farmer. Given that there are 30,000 farms in South Dakota, that comes to billions a year that Lundgren considers unnecessarily spent on pest management in the state.

What got us here? Lundgren points out that "nitrogen fertilizers opened the door for monocultures. Once you go to monocultures, the only way to maintain productivity is with pesticides." And this sets the stage for a treadmill of chemical dependency that allows those selling the chemicals to keep harvesting dollars off a farm—farming the farm, you might say. Just imagine how well a farmer could live on an extra $250,000 a year, not just in South Dakota but anywhere across America. While regenerative methods are not cost-free, the input costs for things like cover crop seeds are nowhere near the outlays needed for conventional methods.

After running through this example, Lundgren relates a video he recently saw of a Pennsylvania farmer saying, "Every year I borrow $800,000 to make $850,000." Just where, Lundgren asks, is that $800,000 going? "The farmer only gets $50,000. When you eliminate the biodiversity of a farm, you have to replace it with something. That biodiversity did something. It was the workhorse of the farm. Lose it and you have to replace it with a jug—and it doesn't work." It also costs a lot.

Back at the Sioux Falls airport we pass a wall-size painting of downy gentian. This gorgeous bit of public art reminded us of Andy Warhol's quip about the beauty in taking care of the land. Perhaps the self-styled provocateur didn't go far enough though. Taking better care of the land would take better care of cows—and us too.

FAT OF THE LAND

Men are not so much the keepers of herds
as herds are the keepers of men.

—HENRY DAVID THOREAU

Looking for milk at the grocery store got a lot more complicated in recent years. When our parents were young, the choice lay between a few brands—if, that is, one could find more than one option. Milk was just milk. It came from a local or regional dairy, and the cows mostly ate grass. When we were growing up in the 1970s, milk from cows still dominated the dairy cases of grocery stores, with options for different amounts of fat. Now, to find whole milk, cream, or half-and-half in the dairy section, you have to wade past all the other "milks"—soy, coconut, hemp, almond, and so on.

Faux dairy products represent the most recent step in a century-long shift away from milk made in cows that ate living plants in a pasture. What did we lose in pivoting away from grass-fed milk? More than advertised, as industrialized animal husbandry optimized production for efficiencies dependent on cheap grain and energy. But it needn't be this way. It's possible to both increase the nutritional quality of milk and meat and reduce the environmental footprint of animal agriculture. We'd just have to raise and feed animals really differently.

Ruminants—cows, sheep, and goats—are long-running mainstays of farming systems around the world. Their moniker comes from the rumen, an underappreciated part of their bodies crucial for their brand of herbivory. The rumen serves us well too, turning grass and leaves we can't digest into meat and milk we can.

The rumen is not a stomach. Those dissolve things. A rumen is an ecosystem that houses most of a ruminant's microbiome. As in any other ecosystem, physical terrain creates different habitats that sustain various inhabitants. Stout ridges of cartilage lend strength and form to the rumen, creating distinct compartments that facilitate the fermenting action and movement of trillions of microbes navigating a sea of gastric juices, saliva, and partially chewed plants. Disturb it or feed it differently, and its tiny inhabitants change what they do and make inside their host. And as goes the rumen, so goes the ruminant—and the nutritional profile of her milk.

The size of a rumen generally scales with the size of the animal. Consider cows. A standard-size beer keg holds about 15 gallons. Add a second keg, and you'd be up to the size of the rumen in a small cow, about 30 gallons. Add a third keg, top it up with a few growlers, and you get to the large end of the rumen range, around 50 gallons.

After a century poking around in the rumens of a wide variety of domestic livestock breeds, scientists still don't fully understand this alive and dynamic place. It's a challenge to study because of the perpetually shifting relationships between microbiota and a ruminant's diet. Nothing stays exactly the same from one day, or season, to the next. How ruminants acquire a good deal of their protein illustrates some of this dynamism. The bodies of bacteria are rich in protein, and so as they die, they flow out of the rumen and break down into amino acids in the lower reaches of the digestive tract. Ruminants use these amino acids to build the proteins in their bodies, including those that become the milk and meat we consume.

Ruminants evolved to use their internal inhabitants to derive energy from living plants, especially young, actively growing vegetation. As we'll see, grazing on healthy forages not only favors cow health but leads to a more favorable nutrient profile for the meat and dairy we eat. But that's not the kind of diet most cows consume—not anymore anyway. People aren't the only animals that now eat a lot more grain than evolution shaped them to.

Fat Rules

Ruminal microbiota will ferment whatever a ruminant eats, creating two main by-products—gases and fatty acids. A ruminant's diet affects both gas production and the types of fats that become part of its milk and meat. The lion's share of the gases are methane and carbon dioxide. If enough accumulate in the rumen, an animal can die. So ruminants employ a simple solution—belching. Fatty acids, the other main by-product of fermentation, leave the rumen and go on to provide the lion's share of a ruminant's energy needs. These same fats become part of our bodies when we eat meat and dairy products.

To understand the implications of these connections for human health, it's helpful to know a little more about fatty acids. While there are loads of details and wrinkles to consider, the essentials of fatty acid biochemistry are fairly simple.* Connect two or more fatty acids together, and it makes a fat. Sometimes a single fatty acid can spur important biological activity in a ruminant. Other times it takes a full-fledged fat molecule. For the most part we'll use the terms interchangeably.

The length of a fatty acid chain ranges from 2 to more than 20

* Fatty acids are shaped like a chain, with carbon atoms forming the "links." A methyl group (CH_3) bookends one end of the chain, and a carboxyl group (COOH) bounds the other end.

carbon atoms. Short-chain ones provide energy to power the basics—turning plants into muscle tissue in meat or fueling a mammary gland to make milk. Ruminal microbiota can transform short-chain fatty acids for their own purposes, generally lengthening them. These longer chains go on to become part of the nutrient package that supports the health of ruminants and those who consume meat and milk.

Fats and their constituent fatty acids are an incredibly diverse class of molecules. The milk of cows, for instance, contains about 400 different types of fatty acids, with about a dozen dominating the fat content of milk.

The biological influence of a fat depends on two fundamental chemical features—the length of its carbon chain and whether or not hydrogen atoms have bonded to the carbons. This second feature corresponds to something you have likely heard of—saturated fats. When hydrogen atoms bond to every single carbon atom along a chain, the fat is considered fully saturated. This makes a saturated fat rather boring, at least chemically speaking. It's simply not very reactive due to lack of spots along the carbon chain for other atoms to drop in and occupy.

Why does this matter? A more reactive fat can do more things, and the most reactive are those that are least saturated. You may know of such fats by another name—polyunsaturated fats, or PUFAs. They occur in both plant and animal foods and include popular cooking oils pressed from the seeds of plants, like safflower, sunflower, and soybeans. Monounsaturated fats, those with just one unfilled spot along a carbon chain, are less reactive compared to PUFAs. Olive oil is this type of fat. We need both saturated and unsaturated fats for optimal health.

In terms of the length of a carbon chain, those that are longest—18 carbons or more—play particularly important roles in human health. Two such varieties of PUFAs—omega-3s and omega-6s—have been the subject of extensive research in connection with inflammation and chronic diseases. We'll return to this topic later, but by virtue of their being PUFAs, you already know they readily interact with other molecules.

The numerical parts of the names omega-3 and omega-6 reflect the first location along the carbon chain free of a hydrogen atom. Such open spots along a carbon chain take on a double bond. An omega-3 fat has such a double bond right after the third carbon atom, counting from the tail end of the hydrocarbon chain. In omega-6 fats, the first double bond comes after the sixth carbon atom. This simple distinction makes a profound difference in what these two types of PUFAs do.

The unflattering lens through which we view fats—the pudgy stuff packed around our middles and on our behinds—belies the prominent roles fats play in our immune, cardiovascular, and nervous systems. The breadth of their biological activity arises from their diverse chemistry, again mainly due to the length of the carbon chain and the degree of saturation.

One of the places this is most apparent is in the membranes enveloping each of the trillions of cells composing the human body. Any given cell membrane faces a big challenge. It needs to be fluid, flexible, and fast-acting so that a wide range of molecules from water to glucose to phytochemicals can get into a cell and waste products can move out. But at the same time a membrane has to be strong enough to keep the contents of a cell from leaking outward. And it is the chemical diversity of naturally occurring fats—both saturated and unsaturated—that underpins how well cell membranes function. Both have their place, but too much or too little of either can undermine health.

The fats found in cell membranes also serve another vital function: they are the supply depot for must-have components that immune cells use to make molecules that regulate inflammation. In addition, fats act like messengers to coordinate and orchestrate things as fundamental as vision and cognition. The way fats function, or fail to, offers one of the clearest examples of how a molecule we get from a food in our diet intimately affects our bodies, minds, and well-being.

What influences the composition of the fat in our bodies? The answer, unsurprisingly, is fats in the animal and plant foods we eat. And their fat composition in turn hinges on what livestock eat and how we grow our crops. For ruminants in particular, the types of plants they eat—dead or alive, leafy or seedy—and how rumen microbiota modify them largely determines the fat profile of the dairy and meat we eat and thus what eventually becomes our tissues. This simple reality connects us all to the land and the health of the soil beneath our crops and pastures.

How the fat of the land becomes you begins with a simple fact. We mammals suck. This is what defines us. It unites people, possums, coyotes, and cows with every other mammal on the planet via our respective moms. Their mammary glands make and deliver fat-rich milk. Mother's milk provides suckling infants with a rich menu of compounds to kick-start life. And nothing churns out fat like a mammary gland.

Genes govern the amount of fat in the milk of mammals. Take arctic-dwelling hooded seals. That their milk is about 60 percent fat is no accident. Fats store a lot of energy, just what warm-blooded seal pups need to grow and thrive in some of the coldest places on Earth.

Small differences in fat content can make a big difference in dairy products. Do yourself the pleasurable favor of conducting a taste test on ice cream made from the milk of a Jersey or Guernsey cow. Both are the smaller-bodied cows that come in fawn, umber, and other earth tones. Their milkfat falls in the upper part of the range for bovines, around 5 percent.

Then try some ice cream or cheese made from the milk of their Holstein sisters. You can't miss a Holstein lady. The largest and most common dairy cows in the United States, they sport coats of snazzy, puzzle-shaped blocks of black or red against a white background. The fat content of Holstein milk is at the lower end for bovines, around 3 percent. This generally translates into less rich and less tasty dairy products than those made from the milk of Jersey or Guernsey cows. That human milkfat also falls in the range of 3 to 5 percent fat helps explain why cows became so embedded across human cultures and diets.

Breeders aren't that interested in the quality (or taste) of ice cream. Their eye is mostly on getting cows to produce a lot of milk. Holstein biology cooperates, shunting energy that would have gone into fat production into making more milk but with less fat. This is why Holsteins are the cows of choice at most large-scale dairy operations. Nothing churns out milk like the udder of the largest dairy cow on Earth.

All in all, ruminant milk is nutrient-dense stuff. Few foods in the human diet come close to its nutritional package. It's a complete protein source, meaning it contains the nine amino acids our bodies must have but cannot synthesize. Milk also contains a significant amount of water-soluble B vitamins, as well as fat-soluble vitamins A and E. As for body-building minerals, milk tends to have a lot of calcium, magnesium, and phosphorus, as well as some zinc and selenium. And just as for grains and vegetables, the vitamins and minerals in milk vary, reflecting soil health and farming practices used on the pastures and crops cows eat. And the fat and phytochemical content of both milk and meat reflect what the cow ate.

Why does all this matter? Because the balance and composition of fats in milk and meat are utterly different when ruminants eat a diet of dead and highly processed plants instead of living plants from a pasture. Seen through this frame, conventional methods of raising and feeding cows affect both animal health and that of the people who eat meat or dairy products.

Cows on Corn

Before the Second World War, dairy cows around the world grazed on pasture as much as weather permitted. They walked out of their milking barn into a pasture and ate living plants, transforming the botanical world into nutrients that nourished them and us as well. At the same time their manure and urine fertilized the land.

Bovine diets changed when people decided to take them off pastures and start feeding them more dead plants and concentrated sources of energy like feeds made from seeds—corn and soybeans in particular. Seeds pack lots of stored energy that helps young plants get off to a running start. This new, high-energy diet goosed milk production and grew meat faster than grazing cows on pastures.

What drove the shift? The combination of cheap fossil fuels and increased grain production made it economical to harvest, process, and transport grains over large distances. By the 1960s American farmers were growing so much corn that it proved cheaper to turn the corn into highly processed carbohydrate-rich cattle feed than to provide animals with access to living or fresh forage. And keeping cows indoors saved farmers the expense and bother of moving them from pasture to milking parlor and back again on a regular basis. Chasing the promise of producing more milk per cow, large indoor feeding operations grew to house dairy herds across Europe and North America.

Breeders also selected for larger cows to maximize milk production per cow. This choice pushed bovine bodies beyond what grazing could sustain. High-production dairies filled with high-performance cows relied on high-octane feed of energy-dense concentrates—mostly corn in the United States and soy in Europe—to keep the factory running.

In contrast, pasture-based dairy farmers favored smaller cows that thrived on grass and other live forages rich in fiber and phytochemicals. Although pastured cows produced less milk, farmers paid less for feed and typically sought to maximize milk production per acre rather than output per cow. A farmer with fertile soil that produced lush foliage could stock more cows and make up for lower production per cow. These two bovine lifestyles—living indoors and eating lots of dead plants and simple sugars versus living outdoors grazing plants in a pasture—led to drastically different outcomes for the soil, the cows, and the fat profile of milk.

Geography too played a role. Cattle and other ruminants are well adapted to temperate zone climates where they can graze year-round—the British Isles, New Zealand, and California all come to mind. In more continental and alpine regions, where year-round grazing isn't viable, animals were moved indoors in the winter and ate conserved forages (like baled hay) until returning to outdoor living in the spring.

Today, the majority of American dairy cows live indoors year-round, with roughly a third living in settings of 500 or more animals. The trend toward large indoor dairies caught on in Europe as well, where pasture feeding has declined over recent decades. In alpine Austria, where half the agricultural land traditionally lay in pasture, more than four out of five dairy cows now live indoors. Less than a fifth even have seasonal access to pasture. All the labor and conflicts involved in moving cows around mountainous lands enhanced the appeal of indoor milk production. Now European countries import millions of tons a year of soy-based feed for cattle that formerly ate living plants. In contrast to the United States and most of Europe, many of the dairy cows in Ireland and New Zealand still consume most of their diet as living plants from pastures.

On grass-fed dairy farms the sun grows the grass, and the cows harvest, process, and deliver it as milk to a central collecting place. In contrast, concentrate feeding requires a lot of fuel and fertilizer to grow, harvest, and process the feed and then serve it to cows confined to indoor facilities. The latter method is more work for us, takes far more fossil fuel, and turns ruminant manure from a free natural fertilizer into noxious waste. Yet so long as corn, soy, and energy remained cheap, concentrate feeding profitably produced the most milk per cow. As progress marched on, cows moved indoors.

The dominant economic factor driving conversion to year-round indoor concentrate feeding was the ratio of the price of milk to the cost of grain-based feed. Few dairy cows remained pastured where milk prices stayed high relative to concentrates, such as in the United States where milk prices were supported and cheap grain subsidized. In contrast, dairy cows tended to remain pastured where milk prices stayed low relative to the cost of grain-based feed. Since the 1950s U.S. agricultural policies sought to maximize grain production and keep the price of commodity corn and soy low to produce cheap, processed foods. The approach worked all too well.

Between 1950 and 2000 average corn yields quadrupled in the United States. What to do with it all? Feed it to cows and people. The same went for soybeans. Soybean oil now accounts for about half the worlds' edible oil consumption. If you fly across the Midwest today, the view consists of mostly corn and soybeans. Unfortunately, devising an inexpensive indoor lifestyle for cows often worsened their health and the healthfulness of dairy products (and meat, as we'll see).

These new problems arose in part because few were interested in questioning what such a dietary and lifestyle shift might mean for rumen dwellers, a key driver of health in all ruminants. Switching out a fiber and phytochemical-rich diet of living plants for a grain-dominated diet leads to pH changes that wreak havoc in the rumen and can kill an animal. It's telling that much of the research and content filling animal science journals these days focuses on ways to address myriad digestive, fertility, and maternal health problems that accompanied the shift to concentrate feeding. The solution proven to work—grazing outdoors on live forages—is rarely featured as a way to fix the identified health problems. Never mind that ruminant biology has been honed, tuned, and tested for about 50 million years, since shortly after the dinosaurs died off.

Living plants became a shrinking portion of what livestock ate as industrialization transformed animal agriculture. Instead of ruminants relying on the wisdom of their bodies to decide what to consume, people tried feeding them all manner of things our big monkey brains cooked up. Would they eat fruit waste like citrus peels, spent brewers' grain, pulp left over from seed-oil production, or even recycled cardboard? Animal scientists tried them all in the search for cheap fodder that produced as much milk and meat as fast as possible.

Today, most American dairy cows eat total mixed rations, or TMRs, the ruminant version of an MRE—the "meal, ready-to-eat" developed for the U.S. Armed Forces. There is an important distinction though. The engineered mixtures of protein, carbohydrate, fat, minerals, and vitamins in bovine TMRs often include appetite stimulators, antibiotics, and growth hormones.

As far as the caloric part of a TMR diet goes, it generally consists of hefty amounts of corn and soy as well as barley and oats, all of which are rich in omega-6 fats.* A diet dominated by TMRs means cows are eating more omega-6 fats than they ever did in the past. And so when we consume their meat and milk, it loads omega-6 fats into our diet, notably to the cell membranes and tissues composing our bodies.

We are solely reliant on the foods in our diet to get both the dominant omega-6 fat in the human diet, linoleic (lin-o-LAY-ick) acid, or LA, and the dominant omega-3 alpha-linolenic (lin-o-LEEN-ick) acid, or ALA. The importance of both these fats earns them a special name among nutrients—essential fats—as our bodies cannot make them. What a ruminant eats is what you become. And it takes both types of fats, not just one or the other, to mount a normal, effective immune response. However, when the diet of cows switched to TMRs,

* Not all seed oils are rich in omega-6 fats. Canola oil has low amounts of omega-6 fats relative to most other seed oils. And the fatty acid composition of linseed (flax) oil is high in omega-3s.

milk and meat became a pipeline for getting far more omega-6 than omega-3 fats into the human diet.

What's in Your Milk?

As mentioned earlier, an animal's genome largely caps the overall amount of fat in their milk. But the types of fats vary quite a bit. Within a breed, milk produced from TMR-fed cows tends to have a lower milkfat content but with a higher proportion of saturated fats than in milk from pasture-fed cows. More fresh grass in a dairy cow's diet increases the percentage of unsaturated fatty acids in her milk, thereby reducing the proportion of saturated fats. And a large body of scientific literature documents that the milk of pasture-grazed, grass-fed cows contains a lot more omega-3s and far less omega-6s than the milk from indoor cows that are fed TMRs.

While animal scientists have long known that a cow's diet influences the fat composition of her milk, one particularly comprehensive and thorough 2013 study focused specifically on dairy cows that ate two different diets. One diet met national standards for organic milk, which require that dairy cows get at least 30 percent of their diet from certified organic pasture, have access to the outdoors year-round, and graze for at least 120 days over a calendar year. Organic milk production allows grain-based feeds (from organically grown crops), and in this study they accounted for about a fifth of the cows' annual diet.

Cows on the conventional diet ate few to no living plants. So although framed around the question of conventional versus organic milk, the study indirectly addressed the difference between a mostly pasture versus a TMR diet.

The researchers conducting the study collected almost 400 milk samples over a year and a half in seven regions across the United States. Just over half the samples were from organic dairies and the remainder from conventional ones. On average, the organic milk had two-thirds more omega-3s and a quarter less omega-6s than the conventional milk.

Would the differences in these fats be greater if dairy cows were to eat even more living plants than required under the national organic standards? This was the question motivating a 2018 follow-up study. It compared the prior study results to the fat composition of more than a thousand samples of "Grassmilk"*—milk from cows that primarily ate living forages. The Grassmilk had one-and-a-half times more omega-3s than organic milk, and you'd have to drink two-and-a-half glasses of conventional milk to get the same amount of omega-3s as from one glass of Grassmilk.

These findings are consistent with other studies investigating how changing a ruminant's diet shifts levels of omega-6 and omega-3 fats in milk. In general, more time spent outdoors foraging in a pasture increases omega-3s. In contrast, increased consumption of corn silage (a mix of dead leaves, stalks, and corn) decreases omega-3s and increases omega-6s, as do grain-based concentrates. A review of 78 studies compared the fatty acid composition of milk from cows with feeding regimens that ranged from year-round grazing to concentrate-based indoor feeding. Researchers found that the basic omega-6 (LA) and omega-3 (ALA) content directly reflected what the cows ate. In other words, the simplest way to increase the omega-3 content of milk and bring it more in balance with historical ratios of omega-6s to omega-3s is to let cows graze.

The difference in the omega-3 levels of milk from a cow on a diet rich in living plants rather than dead plants and grain-based TMRs has to do with the fact that like us, our livestock cannot synthesize the must-have omega-3 and omega-6 fats. So where do they get them?

---

* Grassmilk™ is a brand of milk that specifies cows must consume at least 60 percent of their diet from pasture plants, twice as much as USDA organic standards mandate. The cows used in the 2018 study, however, consumed even more, taking in an almost completely forage-based diet.

From the fat of the land, Earth's leafy green cover. Living plants are the largest, naturally occurring terrestrial source of unsaturated fats. Alpha-linolenic acid, the basic omega-3, makes up more than half the fats in the leaves of herbs and grasses. So when cows chomp, trim, and clip young blades of grass and other leafy plants, they scarf down a lot of ALA.

Omega-3s are integral to photosynthesis, which means they are concentrated in chloroplasts, the special structures in plant cells that pull off the chemical wizardry of converting sunlight into energy. Since chloroplasts reside in leaves, this makes living plants a gold mine of unsaturated fats that regenerates itself after animals nibble and graze.

Seeds are the other major botanical source of fats, but unlike leaves, they are rich in omega-6s and thus a profoundly different kind of food for ruminants. Seeds are built for storage, not for photosynthesis or regrowing rapidly when grazed.

Although there are a number of different kinds of fats in seeds, linoleic acid, the basic omega-6 fat, is abundant. The tissues of ruminants that eat a grain-rich diet marinate in omega-6s. As we'll get into later, a diet packed with omega-6s and light on omega-3s isn't good for cows or us. Ideally, a balance or near-balance of the two is best healthwise. In the research done on Grassmilk the ratio of omega-6 to omega-3 was about 1:1, while organic milk had twice as many omega-6s and conventional milk had almost six times more.

In general, living plants have much higher levels of fats than dry forages like hay or silage. This makes sense. A living plant needs a lot of chloroplasts to generate energy to grow, set fruit, defend itself, and so on. Upon death, a plant's fat-laden chloroplasts break down, and the omega-3s quickly start decaying. Studies comparing wilted and unwilted forage find that dead grass can have a third to a half fewer omega-3s than fresh grass. Conserved forages like hay typically have less than half the omega-3s of fresh forage. This is why cows on a diet flush with green, actively growing leafy parts have more omega-3s and fewer omega-6s in their milk and meat.

Freshly cut, baled plants experience lower losses in omega-3s if they undergo ensilage, a bacterial fermentative process that creates a low oxygen environment in which certain fermenters slow down the decay of plant tissues. This helps to preserve the omega-3s and other nutrients in the plants. Silage made from clover, for example, can have even higher omega-3 levels than living grass. So whether plants are eaten alive or dead, and how they are conserved for later consumption all shape the nutritional profile of milk and dairy products.

Just as for ruminants, our diet shifted to include far more omega-6s than omega-3s after the Second World War. The increase in omega-6s is due largely to food processors favoring seed oils that take longer to go rancid. Spoilage-resistant fats are ideal for processed foods that can sit on a shelf for a long time. So omega-6-rich seed oils in processed foods became an American staple. Just as for white rice and flour, prioritizing storage inadvertently shortchanged nutrition and health.

But don't go thinking it's a good idea to totally strip omega-6s out of your diet—or a ruminant's. Consider that the basic one, linoleic acid, is the precursor for another fatty acid, conjugated linoleic acid (CLA). This particular fat is found only in meat and dairy. That's right, *only* in animal foods. How so? Particular rumen-dwelling bacteria have a lock on CLA production.* And even though the idea that an animal fat would prove beneficial to human health runs counter to conventional thinking about fats, CLA turns out to have anti-inflammatory effects. This takes on special significance given that dairy products provide

* Ruminal microbes make conjugated linoleic acid from linoleic acid (the essential omega-6) through shifting the position of the first double bond along its hydrocarbon chain one position farther along, turning it into an omega-7 fatty acid. This little change makes a big difference, flipping its effects from inflammatory to anti-inflammatory. In addition, some CLA is made in the ruminant mammary gland. Certain gut bacteria in the human microbiome also produce CLA.

about three-quarters of all CLA consumed in the United States. Consider too that between the 1960s and the early 2000s the CLA content of milk from Dutch cows fell by more than half. This means anyone drinking their milk would now have to drink twice as many calories to get the same benefit.

The 2013 comparison study mentioned earlier also found that organic milk averaged almost a fifth more CLA than conventional milk, with CLA half again higher in organic milk during summer grazing. The follow-up 2018 comparison found Grassmilk had more than twice the CLA of conventional milk and over half more than organic milk.

Other studies likewise found more CLA in the milk of pasture-fed cows than milk from cows fed mixtures of grain and silage. Moreover, shifting to grazing can rapidly double the CLA content of milk, or do the opposite in less than a week after switching a cow from pastured grazing to indoor consumption of corn silage.

Numerous studies report seasonal variations in CLA, with higher levels in spring. Why? That's when many grass-fed dairy cows begin grazing on pastures. This isn't surprising as CLA-producing bacteria in the rumen thrive on the fiber in leaves and stems. One study even found that cows grazing pasture and receiving no supplemental feed produced milk with five times the CLA compared to cows fed a typical TMR diet. And because CLA is stable in milk and dairy products, what the cow eats gets passed on to us.

A particularly interesting Swiss study compared the milk quality from—and economics of—cows grazed on lowland pasture versus indoor-fed cows. The milk from the pasture-fed dairy cows had less saturated fat, twice as much omega-3 fat, and three times more CLA. Yet milk from the indoor-fed herd fetched a higher price due to high total fat content and generated greater income due to higher yield. Still, despite the greater income from the indoor-fed cows, it did not cover the cost of machinery, silage, and their cornmeal feed. The lower-cost pasture system proved more profitable. Even without a marketplace premium, it paid to produce grass-fed milk with higher omega-3 and CLA content. Under the modern indoor-fed system, farmers produce higher yields but make less money, even though Swiss consumers pay higher prices for the lower-quality milk.

Interestingly, the creation of CLA arises because of a paradox about the rumen. As much as this ecosystem revolves around and thrives on its bacterial members fermenting the fiber in living plants, a higher intake of LA, ALA, and other types of PUFAs also present a serious challenge. Such fats interfere with a key metabolic pathway that the bacterial fermenters rely on for their own energy, so they change things to their liking through some deft chemistry that happens to generate CLA. This process has big benefits for ruminants as it keeps key members of the rumen community on the job. But should these particular bacterial populations decline precipitously or perish, it changes the rumen pH to resemble a vat of acid, unleashing no end of acute and chronic maladies for the host animal as well as less CLA.*

The typical treatment—killing off rumen microbiota with antibiotics—often exacerbates the problem and leads to chronic acidity, or acidosis, one of the most significant health problems for ruminants. A hopelessly scrambled rumen incapable of robust CLA production means we don't get as much in our milk (or meat).

Disruption of the rumen due to a grain-rich diet also explains other major differences in the nutritional fat profile of grass-fed and grain-fed milk. A European comparison of milk from organic and conventional farms in Denmark, Italy, Sweden, and the United Kingdom found up to twice as much CLA, ALA, and antioxidants (vitamin E and carotenoids) in milk from cows that grazed or ate fresh grass compared to milk from those fed a conventional diet. The study concluded that

* The process of detoxifying excessive PUFAs in the rumen is called biohydrogenation. It involves a most helpful bacterium, *Butyrivibrio fibrisolvens*, adding hydrogens to a PUFA's carbon chain, thereby creating new types of fat in the rumen, one of which is CLA.

while organic or low-input dairy management produced healthier milk, most of the difference was due to greater reliance on fresh forages on the organic farms.

An informative experiment from the Azores archipelago, a thousand miles west of the coast of Portugal, tracked changes in the fatty acid composition of milk produced by Holstein cows that grazed for ten days on pasture with some supplemental concentrates, then switched to a full TMR and concentrate diet for three weeks, and finally switched back to pasture for another three weeks. While the total amount of PUFAs in the milk did not shift much over the course of the experiment, the mix of fats changed a lot. The CLA content dropped by two-thirds during TMR feeding and then returned within a week to initial levels once back on pasture. Similarly, the ratio of omega-6 to omega-3 during the initial pastured feeding was just over 2:1, but climbed above 6:1 during TMR feeding, and then progressively fell back to just under 2:1 once the cows returned to pasture.

In the end, all this biochemistry is pretty simple. Feed a dairy cow forages rich in omega-3s, and she produces milk with more omega-3s.

Butter, Cheese, and Phytochemicals

Cows don't eat what they used to, so milk isn't what it used to be. Neither is butter or cheese.

An Irish study designed experiments to test for differences in butter between pastured, grass-fed dairy cows and those fed a TMR diet and found pasture feeding improved nutritional value. Not only did grass-fed butter contain more than twice the CLA and just half the LA, it also had more beta-carotene and a significantly lower thrombogenic index, a measure of risk for inducing blood clots. Notably, the TMR butter had no omega-3s. None. The grass-fed butter had a substantial amount of omega-3s, as well as preferable color and flavor.

The different milk fat profiles from pasture-fed and TMR-fed cows carry over into cheese as well. In another Irish study comparing pasture versus TMR feeding, researchers found that the different diets affected the nutritional composition, properties, and flavor of cheddar cheese. Making it from the milk of pasture-grazed cows endowed the cheese with more than twice the CLA and beta-carotene than found in cheese made from cows on a TMR diet. Grass-fed cheese also had two to four times more of the basic omega-3 (ALA), two to three times less of the basic omega-6 (LA), and lower atherogenic and thrombogenic indices. The TMR-fed cheddar was paler, and the grass-fed cheese was yellower due to higher beta-carotene content.

Similarly, a study of Italian cheeses produced on farms with a seasonal grazing and confinement system found that pastured, grass-fed milk produced higher amounts of unsaturated fatty acids, including omega-3s and CLA, lower saturated fat levels and atherogenic index, and a yellower color. Not only were pastured cheeses healthier, but trained tasters could reliably discriminate between those made from the milk of grass-fed cows and those from a cow on a TMR diet. In addition, pasture-milk cheese had a preferable texture thought to reflect differences in its fatty acid profile.

Milk from pasture-fed cows also delivers more phytochemicals and vitamins A and E than TMR-fed milk. Grazing on fresh forage has been known to produce high beta-carotene levels in milk since the 1930s. Of course, a complicating factor is that different forages contain different amounts of carotenoids and other phytochemicals. It is notable that grazing increases levels of phytochemicals and vitamin concentrations, which, because of their antioxidant activity, greatly reduce the harmful by-products that arise from oxidative spoilage of fats.

Numerous studies report higher levels of vitamin A and beta-carotene in milk during seasonal grazing than when cows are kept in barns and fed silage or concentrates. Once plants are cut, levels of beta-carotene rapidly decline, with losses of more than 80 percent in some cases. Concentrates are also low in many carotenoids as processing involves heating that degrades phytochemicals.

Terpenes are another kind of phytochemical with antioxidant, anti-inflammatory, and antitumor effects. Levels are more than several times higher in milk and cream from cows raised on diversified pastures than from concentrate-fed cows. And the terpene profile in pasture forage carries through to cheese as well as contributing to the flavor.

Similarly, the amount of phenols in milk directly reflects the amount in a ruminant's feed. Flavonoids, for example, can be ten times higher in milk from cows grazing diverse pastures than from cows fed corn silage and more than twice as high as in milk from cows fed concentrates.

More than taste and flavor suffered when cows moved indoors. We inadvertently shorted our bodies of compounds that support our own health. It also didn't help the health of cows.

Health of Dairy Cows

As dairies shifted to confined settings and a diet of energy-dense rations, high milking performance became the focal point of North American breeders. This changed the bodies of North American dairy cows relative to New Zealand Holsteins in several fundamental ways—body weight, body condition during lactation, and milk yield. North American Holsteins are 100 to 200 pounds heavier, tend toward lower body condition,* and produce more milk with lower percentages of milk fat and protein than their pastured Kiwi cousins.

Lighter-weight cows are better suited for pasture life. For one

* Body condition scores range from a low of 1 for a very thin (starving) cow to 3 for a cow in average condition and 5 for an excessively fat (obese) cow. It is visually assessed based primarily on the condition of the loin and rump, essentially by how much fat—how much stored energy—the cow can draw on to produce milk.

thing, they do less damage to their food supply during wet weather because they don't compact the soil as much. They can also obtain sufficient energy from grazing to maintain their body condition, whereas larger North American Holsteins bred for maximal milk production require so much energy that when grazing on a pasture they can't eat enough plants fast enough to maintain good body condition and health. They've been bred out of their own nature.

Breeding a super-cow that could subsist on a diet of energy-rich concentrates wound up driving the trajectory of the dairy industry. That cows share some of their diet-derived energy with their rumen dwellers and expend energy walking around grazing were seen as wasteful in a system aimed at maximizing milk output per cow. Since 1950 average annual milk production from an American dairy cow tripled. But increased production decreased their fertility, fitness, and longevity.

Even with artificial insemination, infertility has become a major problem in confinement-based dairies. In New York, for example, a state once known for dairying, the average rate for successful impregnation upon first insemination fell by roughly a third since the 1950s. Between 1970 and 2000 the average number of attempts to achieve conception for North Carolina dairy herds jumped from less than two to about three. Another aspect of fertility is the number of pregnancies in a dairy cow's lifetime. By the 1960s the average number for U.S. dairy cows fell below four. By the 1990s it was under three. A comparison of New Zealand grass-fed Holsteins revealed that they maintained higher fertility rates than their North American counterparts. The declining fertility of indoor-living Holsteins fed a TMR diet has been characterized recently as one of the greatest challenges facing the dairy industry. After all, calves are one thing the industry absolutely depends on.

Along with lowering fertility, breeding for high milk production also decreased the longevity of dairy cows. It turns out that cows pushed hard to produce maximum quantities of milk just don't live as long. Dairy cows generally give birth at two years of age. Beyond that the average productive (milking) life span of a dairy cow in the United Kingdom fell from four years in 1990 to three years in 2010. In the United States the average productive life span of a dairy cow in 2010 was a little over two-and-a-half years. In contrast, for the past several decades more than half the pastured dairy cows in New Zealand produced for more than five years. A Danish study of several hundred herds found that the more of their life cows spent on pasture, the longer they lived.

The natural life span of a cow can reach 15 to 20 years. Even though such longevity isn't the norm, when cattle eat a diet of grains and live indoors, we are pushing bovine bodies too hard to sustain good health and end up shortening their lives.

After the shift to grain-based concentrates became the norm, once-rare conditions like chronic acidosis and liver abscesses became common disorders in dairy and beef cattle. Chronic acidosis eats away at the mucosal lining of the rumen, allowing bacteria to enter the cow's blood, spread, and damage their liver and other organs. It can also cause diarrhea and systemic inflammation.

These ailments stem from grain-rich feedlot rations starving fiber-fermenting ruminal bacteria and spurring overgrowth of starch-fermenting bacteria. Altering the ecology of the rumen this way has proven disastrous for bovine health. In addition to the digestive and metabolic disorders feedlot cattle experience, they are also more susceptible to infectious diseases. Yet all these health conditions are easily avoided if we allow ruminants and their microbial partners to eat the diet they evolved to consume.

Recognition that a low-fiber, high-starch diet of concentrates and processed grain promotes chronic acidosis spawned numerous studies of how to reduce symptoms. Ways to prevent this include feeding higher amounts of roughage (more grass) and less thoroughly processed grains or manipulating ruminal microbiota with antibiotics. While researchers focused on supplementing grain-based concentrates to reduce acidosis, few were concerned with the effects of reducing the omega-3 or CLA content of milk. Yet, as we'll see, cows are not the only mammals whose health gets shortchanged on a diet overly dependent on processed grains.

Another serious health problem among dairy cows today is inflammation of the udder, or mastitis. Cows kept indoors are subject to udder injuries and infections uncommon among pastured cows. A North Carolina State University study found that dairy cows kept in confinement had almost twice the incidence of clinical mastitis as pastured cows. The recommended prescription for reducing mastitis for dairy cows is to maximize outdoor grazing. Part of the reason is that a diet of fresh grass delivers higher levels of vitamin E and antioxidants that have been shown to increase a cow's resistance to infections at the root of mastitis.

Putting a cow "out to pasture," the euphemism for end of life, is more than a bit of a misnomer. Lack of pasture, in fact, has become the driving source of lameness, one of the chief reasons for early death among dairy cows. Pastured cows have lower severity of hoof disorders, better mobility, and lower risk of lameness than those confined indoors living on concrete surfaces. Among North American dairy cows, some studies have found that more than a quarter were lame and up to a third at risk of becoming lame. The reason is as simple as it is obvious. Hooves evolved to walk on grass, not concrete.

That pasturing is generally considered beneficial to the health of cows shouldn't come as much of a surprise. Grazing is, after all, what ruminants evolved to do.

And if you think a grass-fed diet influences what gets into meat, you'd be right. The fatty acid composition of beef shifted right along with the Western cow diet. Meat ain't what it used to be either.

WHAT'S IN YOUR BURGER?

All flesh is grass.

—ISAIAH 40:6

We all know it, think we know it, or have heard it for sure: fat is bad. But the notion that we can tar all fats with a negative brush belies what they do inside our bodies. Fats direct and guide critical aspects of mammalian biology, activating or suppressing genes that govern metabolism and immunity. They can influence whether we are happy or sad, thinking clearly or wading through mental fog. And in terms of the fats in meat and dairy, it matters what the animals in the human diet consume—for what they eat, we become.

Although our ancestors ate animals long before modern humans evolved, it was only in the twentieth century that animal fats acquired a bad reputation. This image solidified in the 1950s when a famously self-confident American physiologist noted that an unusual number of healthy, middle-aged men were dropping dead from heart attacks. He pinned blame on saturated animal fats, an idea that remains enshrined in popular wisdom.

After the Second World War, Ancel Keys built a reputation by making the case that consuming animal fats, and cholesterol in particular, caused heart attacks. He didn't set out to do this. It's where the data available at the time focused his view.

Keys made a name for himself conducting studies on food deprivation that helped guide efforts to nourish European populations suffering from wartime food shortages. When the newly formed United Nations Food and Agriculture Organization began working to address global hunger, they recruited Keys to chair prestigious committees on calorie requirements and nutrition.

Serving in this capacity, he learned that Italian men suffered fewer heart attacks than their American counterparts. Intrigued, he compared death rates from heart disease for middle-aged men in six countries and found a strong correlation with the overall proportion of fat in their diet. The more fat that men ate, the more of them died from heart disease. In writing up his findings, Keys made the point that deaths from heart disease declined in Nazi-occupied Norway when dietary fats were scarce. He thought saturated fat consumption led to heart disease and invoked cholesterol as the connection.

But in doing so, Keys overlooked how other dietary factors also varied between the countries in his comparison. Notably, those with the highest rates of heart attacks and fat consumption—the United States, Canada, and Australia—also consumed the most sugar and omega-6-rich seed oils. The two countries with the lowest fat consumption and rates of heart attack were Japan and Italy. Diets in these countries were low in omega-6s but high in other fats. The fish-centric Japanese diet delivered a lot of omega-3s, and Italians ate impressive amounts of olive oil rich in unsaturated omega-9s. Keys didn't much consider these factors once his attention locked on cholesterol as the big fat problem in the American diet. Doctors at the time thought high levels of it caused heart attacks. But in zeroing in on saturated animal fats as the culprit, his hypothesized connection didn't capture the whole story. The mix and levels of unsaturated fats also matter to health. Some are better for us than others.

Keys's authoritative opinions about animal fats serendipitously aligned with other interests. Half a century before, in 1910, soapmaker Procter and Gamble patented a new product made from agricultural waste. This feat of genius transformed trash into a food product. In an agroindustrial version of recycling, they turned oil squeezed from the seeds left after harvesting fluffy, high-value cotton into an artificial edible fat (hydrogenated vegetable oil). At first people were suspicious of the newfangled stuff called Crisco. Cooks had always used animal fats like lard or butter. But the company mounted an aggressive decades-long promotional effort giving away cookbooks featuring their new fat as a key ingredient.

Crisco wasn't the only manufactured fat made from plants to hit our bodies. Shortages of animal fats during the Depression and butter during the Second World War led to another new hydrogenated vegetable oil product—margarine. Initially shaded an unappetizing pale white, producers soon found that dying it buttery yellow greatly increased consumer appeal. The balance of fats in the American diet shifted as omega-6-rich margarine made from processing seed oils replaced omega-3-rich butter and animal-derived saturated fats.* While large food companies pumped more omega-6s into our diet, Keys kept focusing on cholesterol and saturated fats.

He went on to develop the Seven Countries Study, an investigation of heart disease among almost 13,000 men between 40 and 60 years old in the United States, Finland, Greece, Italy, Japan, Netherlands, and Yugoslavia. The largest epidemiological study of its time, it began in 1958 and soon reported that greater saturated fat intake was associated with a higher rate of death due to heart disease.

* Another issue with margarine and similar plant-based butter substitutes is that many include synthesized fats now known to have their own harmful health effects.

Keys believed he had uncovered the smoking gun for why so many American men had been dying from heart attacks. He called his idea the "diet-heart hypothesis," claiming that the saturated fats in butter, meat, and other animal foods raised blood cholesterol levels and thereby caused heart attacks. In 1961, *Time* put him on the cover, proclaiming him person of the year.

Yet the causal link Keys endorsed—cholesterol leads to heart attacks—was not the full story. Looking back, it can seem he reflexively dismissed any thought that animal fats might benefit us. But the analytic tools available in his day were just as culpable. Cholesterol happened to be easily detected and measured in blood samples, so Keys looked at what was illuminated under the intellectual lamppost of his field of study.

Like many influential scientists promoting their own ideas, Keys challenged, sidelined, and overshadowed those who questioned his thinking or brought up results and evidence that didn't fit his hypothesis. Shortly after the initial Seven Countries Study was published, two randomized clinical trials that contradicted the diet-heart hypothesis quietly lapsed into obscurity, languishing unpublished until rediscovered decades later. One, the Sydney Diet Heart Study, took place from 1966 to 1973 and followed several hundred middle-aged men who had recently had heart attacks and found that replacing saturated fats with omega-6-rich vegetable oils lowered cholesterol but increased risk of death from subsequent heart attacks. This didn't square with how the diet-heart hypothesis supposedly worked.

Neither did the Minnesota Coronary Experiment, which ran from 1968 to 1973 and replaced dietary animal fats with omega-6-rich vegetable oil in a randomized cohort of over 9,000 men and women. The group that received vegetable oils had lower cholesterol levels but a higher risk of death from heart disease. Unphased, the diet-heart hypothesis zombied on, forming the basis for dietary recommendations that Americans replace saturated (animal) fats with polyunsaturated fats. In practice this meant that through the closing decades of the twentieth century, omega-6-rich seed oils became embedded in the wide array of processed foods taking over the American diet—and those who ate them.

The story on dietary fat has evolved considerably since Keys's day. Recent research on how fats influence health points beyond specific types of fat to the importance of their overall mix and balance. Yet dietary advice has been slow to recognize more nuanced and complex roles of fats in human health. Indeed, the first ever U.S. Dietary Guidelines, issued in 1980, endorsed replacing animal fats with vegetable oils, enshrining a legacy that continues to muddle how we think about fats and what we eat. Other initially controversial topics in the guidelines, like carbohydrate consumption, are more or less resolved: when possible, we should eat them in whole rather than refined forms. Fats, however, remain one of the least settled topics, with pitched battles still raging over how much and what types Americans should eat. In great part this reflects how such advice gets fashioned between the self-correcting blackboard of science, the shifting winds of politics, and commercial interests working to skew things their own way.

Through all the squabbling over fats, little mention was made of their changing mix in the diet of farm animals, leaving Americans to live with the consequences of this unrecognized experiment. Instead, the simplistic Orwellian characterization stuck—saturated fats bad, unsaturated fats good. Keys and others didn't fully understand that a balance of omega-6 and omega-3 fats helps orchestrate the human immune system's essential process of inflammation to deliver not too much or too little, but just enough at just the right times. Nor, for that matter, did they understand that this required a balance of fats in our food.

We'll revisit this connection in more detail later, but for now consider that linoleic acid, the basic omega-6 dominant in ruminants that eat a grain- and concentrate-rich diet, is the precursor for a longer-chain omega-6 fat, arachidonic acid (ARA). This omega-6 plays several roles, including a really critical one—helping initiate inflammation. The typical American diet today contains 10 to over 20 times more omega-6s than omega-3s. This is roughly 3 to 5 times higher than what is considered beneficial for human health.

Our bodies generally have no problem firing up inflammation. It's failing to turn it off that gets us in trouble. With omega-6 fats swamping their omega-3 cousins over the past half century in the American diet, is it a coincidence that inflammation gone awry is a major factor in maladies linked to excessive inflammation, including heart disease and cancer, which top the list of leading causes of death in the United States? Consider too that a number of debilitating neurodegenerative, reproductive, and metabolic ailments afflicting an awful lot of Americans are rooted in excessive inflammation. Viewed in this light, reducing the ratio of omega-6s to omega-3s in the Western diet may be an important dietary change to help stem the ongoing rise in inflammation-related chronic diseases.

The Matter of Meat

With these things in mind, let's take a look at the fats in animals that become the fats in us.* A 2010 review of seven studies that compared differences in the total fat composition of beef cows on a diet of grasses and other living plants versus a TMR diet found no consistent difference in the total amount of saturated fats. But in more than half the comparisons cows on a TMR diet had higher levels of the types of saturated fatty acids that raise total cholesterol levels. Moreover, saturated fats are not monolithic. The grass-fed beef had more stearic acid, a type of cholesterol-neutral fat, while grain-fed beef had more myristic

* Most of the research on ruminants discussed in this chapter applies to beef cattle. However, it is also relevant for other ruminants like sheep and goats.

acid and palmitic acid, which raise levels of LDL cholesterol, the "bad" kind. There were other striking differences too. Meat from grass-fed cows had half more to five times more omega-3s and up to twice the CLA as the grain-fed beef. Averaged across the studies, the ratio of omega-6 to omega-3 for grass-fed beef was less than 2:1, whereas beef from grain-fed cattle had ratios of between 8:1 and 10:1.

The differences weren't just in fats. For the studies that reported values, grass-fed beef had half more to five times more vitamin E and four to more than ten times the beta-carotene than grain-fed beef. This is not an isolated finding. Additional studies also report that the meat of pasture-fed steers contains several times the beta-carotene and vitamin E found in meat from grain-fed steers.

A similar review published the previous year compared the fatty acid profiles of concentrate-fed cattle in Spain and the United Kingdom with those of entirely grass-fed cattle from Uruguay. The intramuscular fat from concentrate-fed cows had omega-6 to omega-3 ratios between 8:1 and 15:1. The grass-fed cows had ratios of less than 2:1. Other studies likewise report that grass-fed beef has up to several times more CLA and much lower ratios of omega-6 to omega-3.

Similar to the influence of a dairy cow's diet on the fatty acid profile of her milk, the types of fat in the diet of a beef cow carry over to their meat. The ratio of omega-6s to omega-3s in grass and legume forages is typically around 1:1, whereas grain-based feeds have omega-6 levels 10 to more than 40 times higher than omega-3s. Simply put, the mix of fats in what cows eat is the mix of fats that build their bodies—and ours.

But while consumers might assume organic means grass-fed, recall that USDA organic standards only require that a ruminant's diet contain at least 30 percent dry matter from organic-certified pasture. In other words, the USDA's organic certification doesn't address what actually composes a cow's diet, only that the crops used as feed are organically grown. So dead plants like silage or grain-based feeds can still make up the majority of a cow's diet. Moreover, ruminants that become the meat we eat are exempt from the 30 percent pasture requirement for the last fifth of their lives, typically their last several months. While animal welfare activists are upset about USDA organic standards allowing confined feeding operations, everyone choosing to eat beef should care about what the cows actually got to eat.

There are various labeling programs for grass-fed meat (and milk), but the definition of "grass-fed" varies. Due, in part, to such confusion, the USDA dropped its grass-fed labeling program in 2016. Originally established a decade earlier, its grass-fed standards were criticized for allowing indoor hay and silage feeding for part of the year and permitting routine additions of hormones and antibiotics to feed. Some private entities now offer strict 100 percent grass-fed certifications.

Yet all young cows eat grass for some time, even if destined for a feedlot. What's of central concern to those who eat meat is knowing what cows ate in the month or two before slaughter, the so-called "finishing diet." For beef cattle, including those that meet organic standards, corn-based finishing is allowed. However, this shifts the balance of fats toward the omega-6 end of the spectrum just in time to shape what lands on our plates.

A three-year experiment on Angus steers compared the effects of a corn-silage and concentrate diet with reversion to grazing for a little over a month before slaughter. While corn-finished cattle had a ratio of omega-6 to omega-3 greater than 6:1, forage-finished cattle had a ratio of just over 1:1. A 2017 Utah State University study similarly found that grain-finished beef had a ratio of omega-6 to omega-3 twice that of legume- or grass-finished beef.

Another Utah State study compared the fatty acid profile for rib-eye steaks from Angus steers raised at their research facility on three different finishing diets—grass, legumes, and grain. Not surprisingly, they found the highest ratios of omega-6 to omega-3 in the group on a grain-finished diet. In addition, the researchers compared two other ribeyes—organic-certified, grass-fed purchased from a largely organic national grocery chain and USDA Choice feedlot-finished purchased at a big-box store—to the three finishing diets (grass, legume, or grain). The grass and legume–finished and organic-certified, grass-fed steaks had ratios of omega-6 to omega-3 of between 2:1 and 4:1, whereas the grain-finished steaks from the research facility had values of almost 6:1. At over 15:1, the feedlot-finished USDA Choice steaks from the big-box store had the highest levels of omega-6s.

A 2016 meta-analysis of compositional differences between organic and conventional meat found that organic meat had almost half again more omega-3s. The report attributed this difference to more grazing and forage-based diets under organic practices. Still, to meet organic certification a rancher need only provide about a third of a herd's diet through living forages.

While there are clear and consistent results from studies comparing ratios of omega-6 to omega-3 from completely grass-fed beef to ratios from TMR-fed beef, studies on "grass-finished" reveal how murky practices are in the area of grass-fed standards. Researchers at Michigan State University analyzed 740 samples of strip loin taken from a dozen farms across ten states and marketed under a grass-fed label. They also surveyed the ranchers about the finishing diet their cows ate in the two months before slaughter. It was highly variable, ranging from fresh pasture to a mix of dried grass and omega-6-rich soybean hulls. In other words, some of the cattle ate a lot of living plants, whereas others mostly ate dead plants.

Differences in the ratios of omega-6 to omega-3 tracked with the fat profile of the feed sources, yet consumers would never have known this from the grass-fed label. While the overall average ratio of omega-6 to omega-3 was 10:1, some of the meat samples had much lower or higher values. Two of the three producers whose beef had ratios of omega-6 to omega-3 greater than 15:1 reported feeding their cows stored forages and grain-based supplements. The cattle from two other producers with intermediate ratios (close to 4:1) received silage as a finishing diet. Of the five producers whose beef had ratios of about 2:1 or less, all reported their cows consumed only living pasture or fresh forage. The beef from producers with the highest ratios of omega-6 to omega-3 also had the lowest levels of beta-carotene and vitamin E, with two of them having no detectable beta-carotene at all. Antioxidant phytochemicals were highest in beef from producers who reported finishing their cattle on fresh forages.

In the end, what this study and dozens of others show is that no matter what we choose to call the diet of a beef cow or the burger on the grill, the fatty acid profile of meat reflects what we choose to feed cattle. And as we'll see shortly, cattle fare no better than people on a diet rich in omega-6 fats.

Mystery Fat

Part of the reason that the essential omega-6 and omega-3 fatty acids, LA and ALA, respectively, are the subject of so much research is that they readily become longer-chain fatty acids that play vital roles in our physiology and health. To tell this story, we'll start with the research of two Danish physicians, Hans Olaf Bang and Jørn Dyerberg.

In the late 1960s, they set out to investigate the high rate of heart disease among Danes. Bang had come across intriguing research suggesting that Indigenous Greenlanders, the Inuit, had remarkably little heart disease. Dyerberg helped him design a study to investigate the difference between the two populations.

The pair scoured local medical records and confirmed that the Inuit rarely died from heart disease relative to study subjects living in Denmark. Then they went to painstaking lengths to collect and preserve Inuit blood samples so they could analyze their fatty acids. Although some challenged the accuracy of the death records used to gauge cardiovascular mortality among the Greenland Inuit, Bang and Dyerberg also collected dietary information from the Inuit on a weekly basis and found that their meals contained extraordinarily high amounts of animal fat from seals and whales, with lesser amounts from fish. Although the sample size was small, the doctors were puzzled. Everyone who studied heart disease at the time knew of Keys and his widely accepted diet-heart hypothesis. Given the amount of fat Inuits ate, heart disease should have been a major killer. But it wasn't.

Bang and Dyerberg faced another puzzle as they examined the fatty acid results. The Inuit had seven times more of one fat than the comparison group in Denmark. Moreover, neither Bang nor Dyerberg had ever encountered the mystery fat in blood samples before. Eager to learn more about it, they sent a sample off to a scientist in the United States with state-of-the-art equipment who identified the new fat as eicosapentaenoic acid, or EPA for short. It was a 20-carbon, long-chain omega-3. It occurs in this form in cold-water fish and can also be made from ALA, the slightly shorter omega-3.

The pair of Danish doctors found something else intriguing as well. The Greenland Inuit had much less of another fatty acid, only about one-seventh the level of study subjects in Denmark. Bang and Dyerberg had encountered this fat before: it was ARA, the omega-6 central to initiating inflammation.

The relevance of what Bang and Dyerberg learned—that the long-chain omega-3 EPA was high in the Inuit, while the omega-6 ARA was high in the Danes—relates to a condition underlying heart disease. Though it wasn't commonly known at the time, chronic inflammation in a person's arteries is a major risk factor for high blood pressure, heart attacks, and other manifestations of cardiovascular disease. Another unknown factor at the time was that EPA ramps down inflammation. In hindsight, the net effect of a high level of ARA combined with a low level of EPA in the subjects living in Denmark meant inflammation was poorly controlled, as if there were no off switch.

Eicosapentaenoic acid, it turns out, has two similar-acting long-chain omega-3 sister molecules also made from ALA. As we'll get into shortly, people can convert some dietary ALA into these long-chain omega-3s. But we are not nearly as good at making them as seals, whales, and cold-water fish.

While EPA weighs in at 20 carbons long, docosahexaenoic acid, or DHA, bests its omega-3 sister at 22 carbons. This makes DHA highly interactive and a major player in some of the most rapid activity in the human body, like nerve cells in your brain that fire a hundred times a second. From womb to tomb, DHA enables our keen sense of sight, maintaining the shape of photoreceptor cells in the retina that respond to light and send the signal on to our brain for processing. If you have ever taken a fish oil supplement, chances are good that DHA and EPA were listed on the ingredient label. Cold-water fish like salmon and mackerel are rich sources of both.

Docosapentaenoic acid, or DPA, completes the trio of long-chain omega-3s. Like DHA, it is a 22-carbon chain and quite reactive.* But DPA is the least studied of the trio partly because a pure form of it is really hard to come by due to its propensity to morph into either of its sister long-chain omega-3s. Even though this trio of omega-3s are called "very-long-chain omega-3s," for simplicity we'll refer to them mostly as long-chain omega-3s but call out individual ones by their acronym as needed.

As part of their research on the traditional Inuit diet, Bang and Dyerberg documented what their study subjects ate. On a weekly basis seal and whale meat (and blubber) each made up roughly six meals, whereas fish accounted for just one or two. While much is made of the beneficial health effects of omega-3 fats found in cold-water fish, such fish were not a major part of the diet for the Inuit population that Bang and Dyerberg studied. Marine mammal fat rich in DPA composed the bulk of their dietary fat.

* Even though DPA and DHA have the same number of carbon atoms (22) they differ in their number of double bonds, with DPA having 5 and DHA having 6.

We, like most Americans, will never routinely consume seal meat or whale fat, much less at the historic levels of the Greenland Inuit. While wild cold-water fish are known for high levels of the other two long-chain omega-3s (DHA and EPA), most Americans simply do not eat much of these fish on a regular basis.

So where can we look for other dietary sources of long-chain omega-3s, especially DPA, the one among the trio that can rapidly transform into either of its sisters? To the landlubber cousins of seals and whales—cows and other ruminants—depending on how they are raised. The fact that our bodies can convert DPA into its sister fats means that meat and dairy from grass-fed ruminants are an underappreciated source of long-chain omega-3s.

Consider, for example, the 100 percent grass-fed cattle that graze on Gabe Brown's ranch, one of the regenerative farms included in the national comparison of regenerative and conventional crops discussed earlier. Brown also shared a lab analysis with us that showed a seven-ounce ribeye steak from one of his steers delivered more ALA (the essential omega-3 precursor for longer omega-3s) than 48 out of 56 species of fish filets tested in a 2014 study of commercially available filets from U.S. seafood vendors, including wild salmon. Brown's ribeye delivered about a third of the daily recommended long-chain omega-3 intake. In contrast, conventional American feedlot beef has been reported to utterly lack DPA.

Brown also shared the results of a regional comparison of the fatty acid profile of beef, chicken, and pork raised on his ranch with equivalent conventional cuts purchased from a local supermarket. The Brown's Ranch meats were entirely pasture raised on healthy soils and had three to ten times more ALA than their conventional counterparts.*

In terms of the long-chain omega-3s EPA, DHA, and DPA, the

* When he started farming in the early 1990s, Brown's soil had less than 2 percent organic matter. Since then, his regenerative grazing practices have nearly tripled it, bringing his soil back to about the level of the native prairie in the region.

Brown's Ranch meat was, respectively, 38 to 91 percent higher, 73 to 350 percent higher, and 58 to 218 percent higher than the conventional cuts. As far as the ratio of omega-6 to omega-3, Brown's Ranch beef was just over 1:1, while the conventional beef was about 6:1. For both chicken and pork the ratio was more than twice as high for the conventional meat. Lastly, the CLA content of Brown's Ranch beef was three times higher than that of the conventional beef.

Numerous independent studies confirm that meat from beef cattle fed a grass-dominated diet differs from meat from those fed a grain-dominated diet. In the 2010 comparison of seven studies discussed earlier, the amount of EPA ranged from about two to more than ten times higher in grass-fed cattle. Grass-fed beef had 20 percent more to over four times the DPA and up to three times the DHA of grain-fed beef. Several years later a review of the effect of grass- versus grain-finishing on U.S. beef reported that grass-finished steaks had a third more to half more LA, two to four times more ALA, and about three-quarters more long-chain omega-3s. And while even grass-fed cattle don't have much DHA, they do have a lot of DPA, the same fat so abundant in the seals and whales of the traditional Inuit diet.

These diet-related differences are not only in beef cattle. Recall the study on milk from dairy cows on three different diets. Grassmilk, from 100 percent grass-fed cows, had half more to two-thirds more EPA and DPA than conventional milk and about a quarter more than in organic milk. As for DHA, only the Grassmilk had detectable amounts.

Despite the potential for grass-fed beef and dairy to increase long-chain omega-3s in the American diet, few seem aware of this potential, let alone recommend it in dietary guidance. Instead, we are urged to eat fish. Yet some of the typical fish in the American diet have higher ratios of omega-6 to omega-3 than grass-fed beef. One study found that the ratio of omega-6 to omega-3 averaged more than 6:1 for half a dozen fish common in the American diet. One, farmed catfish, had a ratio of well over 10:1—a level comparable to grain-fed feedlot beef.

Much of the fish that Americans eat is farmed, even salmon. And just as with cattle, what salmon eat governs their fatty acid profile. Farmed salmon confined in marine pens eat manufactured feeds, whereas wild salmon forage on omega-3-rich food sources like aquatic insects in streams and small fish and zooplankton in the open ocean. Explosive growth in global aquaculture led to replacing fishmeal and fish oil in the diet of farmed salmon with seed oils from land plants, raising the ratio of omega-6 to omega-3 in their tissues to about twice that of wild salmon. A study that compared the fatty acid composition of farmed Atlantic salmon from 2006 and 2015 found long-chain omega-3s fell by half due to changes in feed composition. It seems that today you'd have to eat twice as much farmed salmon to get the same amount of long-chain omega-3s as a few decades ago.

Blood levels of EPA and DHA in populations around the world correspond to dietary intake, with the Japanese, Koreans, and Scandinavians who consume large amounts of cold-water fish having the highest levels. Most of us, however, eat just a third of the minimum dietary recommendation for long-chain omega-3s.

So what about eating more wild fish? This isn't really a global solution as world fisheries plateaued several decades ago and there are only so many fish left to eat or feed to other fish we then eat. But there is another globally abundant source of omega-3s—grasses and other plants that cows and other ruminants evolved to eat.*

A diet of living plants for dairy and beef cows matters because it provides ALA, the omega-3 that humans can convert into long-chain omega-3s, notably DHA, the main omega-3 in fish oil. Adults can convert about 5 percent of their ALA intake into DHA, while preg-

* We are not advocating that forests be converted into grassland or pasture, but rather that grassland environments provide natural opportunities for regenerative grazing.

nant women do considerably better, up to 20 percent, which is particularly important during pregnancy and nursing to support normal growth and development of a fetus and infant. Although some say that these conversion rates are too low to matter, we need all the long-chain omega 3s we can get in our diet. An obvious start down this path is simple: eat fewer foods with lots of omega-6s. For as it turns out, consuming a lot of omega-6s relative to omega-3s reduces the efficiency of converting ALA to long-chain omega-3s. In addition, since grass-fed meat and dairy are sources of long-chain omega-3s, eating more of them is also a way to boost omega-3 intake.

Recall that DPA is the one of the trio that readily converts into the other two—EPA and DHA. Unfortunately, dietary recommendations do not (yet) take into account DPA levels. But perhaps they should as almost half of Western dietary intake comes from meat, with DPA accounting for more than half of the long-chain omega-3s in beef, lamb, and pork.

How much difference might it make if we replaced grain-fed meat with grass-fed? Researchers at the University of Ulster set up a month-long, double-blind, randomized study involving 40 healthy men and women. Study volunteers were asked to forgo consuming oily fish and replace part of their weekly meat intake with one serving of minced beef, one of steak, and four small pieces of lamb. The source of the meat was from animals whose diet was carefully monitored, and for the six weeks prior to slaughter they ate either a fresh grass or concentrate diet. For the study subjects that ate grass-fed meat, blood plasma concentrations of ALA and all three long-chain omega-3s increased significantly (with DHA almost doubling). In contrast, levels of long-chain omega-3s dropped for the group that ate meat from grain-fed animals. As a result, the average ratio of omega-6 to omega-3 in the blood of those consuming the grass-fed meat fell from 9:1 to 6:1, whereas the average ratio for those eating concentrate-fed meat rose from 8:1 to 13:1. The study showed that consuming grass-fed meat could substantially lower the ratio of omega-6 to omega-3 fats in people.

Let Them Eat Grass

Just as for cows, what we feed chickens affects their health, as well as their eggs. Changing the way we raise chickens changed how fat they are and thus how much fat they deliver in our diet. Most noticeably, the total fat content of chicken meat increased dramatically while the essential fat content (omega-6s and omega-3s) declined. That's the conclusion of a London Metropolitan University study that compared data reported in studies from 1870 through 2004. In Britain and America, a typical three-and-a-half-ounce chicken breast had 2 to 4 grams of fat in the late nineteenth century. By the early twenty-first century that same serving delivered more than 20 grams—five to ten times more fat. While the leaner nineteenth-century birds delivered three times more energy in protein, the twenty-first-century birds delivered three times more energy as fat.*

Once a high-protein food, it seems chicken has become a high-fat food. Ironically, a big part of the reason for dietary recommendations to eat less red meat and more chicken came from viewing chicken as a low-fat alternative to beef.

It wasn't just the total fat content that changed. So did the type of fats. At least that's according to a comparison of fatty acid profiles of conventionally raised chickens sampled in 1970 and 2006. The amount of the basic omega-6, LA, increased by almost half, whereas the amount of the long-chain omega-3 DHA dropped to about a fifth

* While critics are quick to invoke changes in analytical methods as a reason to question such historical comparisons, assessing total fat and protein content does not require sophisticated analyses. Indeed, traditional analytical methods are about as accurate as those of today: twentieth-century advances in measurement technology mostly affected the ability to resolve microgram-level differences and allowed automation to process far more samples much more quickly.

of former values, which previously had been pretty close to those for wild birds. So if you were looking to today's chickens as a source of DHA, you'd need to eat a lot more bird than you did in the 1970s. Even organic chicken purchased from British farms and supermarkets between 2004 and 2008 had nine times more omega-6s than omega-3s.

What drove these changes? Just as for cows, the fowl diet shifted dramatically as intensification and industrialization of poultry rearing in the late twentieth century focused on maximizing meat and egg production. Breeding programs accompanied the switch from free-range to indoor caged systems so as to minimize pesky chicken behavior like running around foraging. Birds needed to settle down in tight quarters and get along on a diet of grain-sourced, high-calorie, protein-rich concentrates and routine use of antibiotics to spur growth. Such dietary and lifestyle changes rippled through to a hen's eggs as well.

In 1989 the *New England Journal of Medicine* published a striking example that found substantial differences between the fatty acid composition of egg yolks purchased from an American supermarket and those from a Greek farm where chickens roamed and ate a diet of grasses, insects, dried figs, and a lot of purslane (a fleshy plant rich in omega-3s). The Greek eggs had 10 times more omega-3s, and the American eggs had almost twice as many omega-6s. Subsequent studies confirmed that conventional chickens eating kibble-like pellets made from omega-6-rich seeds and oils tend to produce eggs that have 10 to 20 times more omega-6s than omega-3s, whereas pastured hens that eat insects, grasses, and other live pasture plants generally lay eggs with ratios of omega-6 to omega-3 under 5:1.

A hen's living conditions also affect the fatty acid composition of her eggs. This finding produced friction and debate in some quarters over just what "free range" actually means. Are birds spending significant time outside with good foraging opportunities for plants and real, live bugs? Or do they simply have access to a small patch of bare dirt outside a crowded barn where they tussle over processed food pellets?

Over decades, studies examining the question concluded that spending more time outside foraging improves the fat, phytochemical, and vitamin profile of both chicken and egg. This is not news. It has been known for decades.

Along with fat composition, it is well recognized that a hen's breed and diet govern the vitamin content of her eggs, with vitamin-rich feeds yielding vitamin-rich eggs. And phytochemicals in what chickens eat transfer into egg yolks, putting either an antioxidant-rich or antioxidant-poor egg on your breakfast plate. In particular, a hen's diet influences the amount and type of carotenoids in her eggs, with those eating grass or silage rich in carotenoids laying eggs with notably darker yellow to reddish yolks.

Diet also influences levels of phytochemicals in chicken meat. For example, a 2016 Italian study found significant differences in phytochemical content between the meat of uncaged indoor chickens eating conventional feed and others housed in a similar pen but with access to outdoor forage. The greater phytochemical content in the diet of the outdoor grass- and insect-eating birds translated into five times higher plasma antioxidant concentrations (as well as twice the omega-3s and less omega-6s) than in the meat of the indoor birds.

When chickens and cows forage outdoors, their bodies can synthesize vitamin D, which gets incorporated into eggs and milk. It's a pretty direct relationship. A 2014 German study reported that yolks from the eggs of outdoor-roaming, free-range hens contained three to four times the vitamin D of conventionally produced indoor eggs. In another study, this time on cows, German researchers covered them with horse blankets for a month during summer grazing. Vitamin D levels in their blood tracked the degree of hide covering, increasing with greater exposure to sunlight and decreasing with less. This helps explain why seasonal variation in sunlight typically produces several-fold variations in the vitamin D content of cow's milk.

Interestingly, chickens and their eggs were once a primary source of land-based DHA, but only because they ate a diet rich in the essential omega-3 (ALA), from which they could make DHA. If we want chicken meat and eggs to once again be a source of health-imbuing, long-chain omega-3s in the human diet, there is an easy fix: let them live like chickens and forage in environments with omega-3-rich foods.

And this brings us to another aspect of why it matters what we feed animals that become part of the human diet. The settings in which we raise them affect their health. Put simply, free-ranging animals that forage to feed themselves are healthier. Hens with higher omega-3 levels have higher levels of antibodies and antioxidants, as well as better immune system response and inflammation regulation.

Feedlot Diseases

Like chickens, cattle do not fare well in confinement. Back when Columbus introduced cattle to the New World on his second voyage in 1493, they grazed year-round when possible and ate a variety of conserved forages in regions with severe winters. Centuries later, as tractors replaced animal labor, the practice of growing crops and raising animals on the same farm gave way to larger operations that specialized in one or the other. Subsidies and policies aimed at maximizing production encouraged American farmers to choose between raising grain or livestock.

As farming "progressed," trainloads of grain filled feedlots where young beef cows were sent to live out the rest of their lives. Through the 1950s feedlots ratcheted up the amount of corn in cattle rations. Some even began plumping up steers on an all-corn diet. Large-scale irrigated corn production made for a lot of cheap cattle fodder, and confined feeding operations took root across the country. Today, around 90 percent of all calves end up in a feedlot. Some argue that this is an efficient way to raise cattle as it brings large numbers of cows and calorie-rich food sources together in one location. But this strategy undeniably compromises bovine health.

Bovine respiratory disease (BRD) is the leading cause of sickness and death among U.S. cattle and one of the biggest challenges for feedlot managers. First noticed in the late 1880s, BRD has increasingly plagued feedlots in recent decades despite substantial, ongoing efforts to treat it. Most cattle in the United States reside in feedlots, and a study of bovine health from 1979 to 1994 estimated that BRD contributed to over half of the mortality in feedlots. By 2017 BRD affected about one in six cows in feedlots. Were any one human disease responsible for more than half of American deaths, it would garner national attention and action.

Efforts to prevent BRD through vaccination have proved equivocal at best, and the antibiotics commonly administered to treat it don't work in about a third of cases. Nor have dietary supplements made significant headway. In some feedlots wholesale treatment with antibiotics actually creates more health problems, including higher risk of drug-resistant bacterial infections. Moreover, BRD does not appear contagious, as risk of contracting it does not increase for newly arrived calves when housed with sick cattle. Instead, BRD behaves more like a chronic disease that compromises overall health, including the ability to resist infections.

Diet-related research on BRD in feedlot cattle reveals some interesting trends. It has been known since the 1980s that more calves die on the common feedlot diet of dried cornstalks than on a diet of fresh hay. High levels of grain in the diet of new arrivals are also associated with feedlot mortality due to BRD.

Trace mineral analyses of livestock fodder from 18 states found more than half deficient in zinc and copper, minerals critical for proper immune system functioning. To some extent micronutrient supplements (vitamin E, copper, selenium, and zinc) can compensate for these deficiencies and help decrease BRD morbidity. But the range of chronic and acute health problems that feedlot cattle experience suggests that something is undermining their immune systems.

Perhaps that something is linked to the dramatic change in the ratio of omega-6 to omega-3 fats in their new diet when calves arrive at a feedlot. As mentioned earlier, omega-6s are part of the process that initiates inflammation in people, and this is the case for cattle too. The acute phase of a normal immune response is usually short acting. Once first responder–type immune cells begin to get the upper hand, other immune cells start to churn out compounds that quell inflammation. Omega-3s are a structural part of such compounds, and without sufficient levels, inflammation cannot end like it should. But, as we've described, the typical feedlot diet supplies few omega-3s, compromising a cow's ability to end inflammation. A 2020 study supports this connection in finding that calves from the same herd arriving at a Mississippi feedlot with higher levels of inflammation-resolving compounds made from omega-3s remained healthy, whereas those that developed BRD had lower levels of the same compounds.

Major diseases rooted in failure to resolve inflammation began plaguing cows once we replaced their diet of living forages with one rich in omega-6s. Sound familiar? It's a lot like what happened as people adopted the modern Western diet.

Dairy cows that live in settings similar to those of beef cows are also burdened with ailments rooted in a dysfunctional immune system. After calving, dairy cows mount an intense inflammatory response to bacterial invasion of the uterus. Normally, the inflammation subsides to more normal levels after five or six weeks. But in a quarter to more than half of Holstein dairy cows, uterine inflammation persists to the time of next insemination. This substantially reduces the odds of impregnation, decreasing the fertility of modern dairy cows.

An impaired inflammatory response also appears to underlie chronic mastitis among dairy cows, most likely owing to lack of inflammation-resolving compounds made from omega-3s. So the three big bovine problems—BRD, reproductive decline, and mastitis—all appear to be rooted in the inability of cows to shut off an inflammatory response.

Three other common diseases among feedlot cattle—ruminal acidosis, bloat, and liver abscesses—echo what occurs among confined dairy herds. Shifting beef cows from a forage-based diet to grains or grain-rich concentrates has greatly increased the prevalence of acidosis. It is now recognized as a common disorder among animals in feedlots. Indeed, a 2007 review of bovine acidosis concluded that a typical feedlot diet and living conditions led to acidosis. Ingesting large amounts of rapidly fermentable grains destabilizes ruminal microbes, leading to bloat that can cascade into serious health problems. Liver function also fares poorly on a feedlot diet, with up to a third of cows in feedlots developing abscesses.

Most beef cattle in feedlots don't live long enough to suffer the full effects of ruminal acidosis and abscessed livers the way longer-lived dairy cows do. Instead, BRD is their main nemesis. Feedlot managers obviously prefer cows be free of BRD as a chronic pro-inflammatory response repurposes energy from muscle growth, reducing cattle weight and marbling. And since antibiotics do sometimes work as a BRD treatment, they are commonly administered as a preventive measure. But this leads to other issues. The community of rumen microbiota changes dramatically, which inadvertently reduces CLA production and helps fuel the rise of antibiotic resistance, neither of which helps our health either.*

Yet we know of a way to greatly decrease or altogether prevent the constellation of ailments and misery that can arise from a feedlot diet and lifestyle—grazing on living plants. After all, this is the thing cows and other ruminants know how to do really well. Seeing inflammation

* More than half of the medically important antibiotics sold in the United States are used to hasten growth and prevent infections in livestock.

as something bovine biology can regulate with an adequate supply of dietary omega-3s, rather than something for veterinarians to suppress, would refocus attention on the underlying problem—what we make them eat. As it turns out, when given the opportunity, ruminants know perfectly well how to choose a diet that is good for them—and us.

BODY WISDOM

I'm just an animal looking for a home.

—TALKING HEADS

Choosing what to eat and then preparing and eating it occupies a good deal of our time. We marvel at a falcon, leopard, or spider that does the same. Yet most of us look at ruminants with no such awe and little curiosity. Uninspired, we yawn and move on. Honestly, how much brainpower or prowess could it possibly take to walk around and eat plants? Plenty, as it turns out.

Herbivores are neither stupid nor lazy. In the wild they may encounter over a hundred plant species in a typical day, choosing 3 to 5 for the bulk of any given meal while nibbling smaller amounts of 50 or more. Their skill lies in harvesting nutrients from the botanical world while simultaneously monitoring their intake of plant compounds that could make them sick. This is no small feat. It requires the whole of an herbivore's body—rumen, brain, and gut receptors.

Letting domesticated livestock nourish themselves through grazing and browsing generally makes animal nutritionists uncomfortable given the certainty that an animal will take in some amount of plant-made toxins. Ecologists have a different take on the herbivore lifestyle. From bovines to beetles, they work out how animals dodge phytochemical bombs and shape the composition of plant communities. Molecular biologists can fill you in on the minute details of such relationships. And chemists will tell you about the rings and strings of carbon bound up with various other atoms that produce the deterrent power of any given phytochemicals. However you look at it, herbivores are experts at navigating the gastronomic gauntlet of the botanical world.

Grass and Cow

It is not easy to think like a cow, but almost a century ago a curious Frenchman rose to the task. André Voisin, a biochemist, inspired some of the key insights behind today's methods of regenerative ranching.

Voisin was born in 1903 on a family estate in upper Normandy. After completing a biochemistry degree in Paris, he worked as an engineer in the rubber industry until the Second World War, when he fought with Free French forces and was awarded the Croix de Guerre for heroism in combat. After the war he taught biochemistry at the National Veterinary School in Paris. But his real passion lay in watching cows graze their pastures. Focusing his observational skills and biochemistry background on the relationship between cows and plants, he came to challenge the wisdom of conventional grazing practices.

That forage quality influenced livestock health was common knowledge among farmers in his day. Yet Voisin puzzled over how agricultural researchers routinely assessed potential forages without accounting for the behavior of animals actually walking around pastures choosing what to eat. Instead, researchers served cows precut forages in tidy, controlled feedlots that emulated the growing trend of moving cows indoors. Textbooks on animal nutrition at the time extensively covered the nuances of a feedlot diet, but Voisin complained that they devoted few words to grazing practices, even though this was how cows had eaten for centuries. Time and again experts designing feedlot rations were surprised when cows did not show enthusiasm for their concoctions. Voisin focused instead on how cows approached eating, on the relationship between animal and pasture—the meeting of grass and cow.

In his 1957 book *Grass Productivity*, he advocated for what he called rational grazing, a system based on frequently moving cows among paddocks, to allow adequate time for the grass in each field to recover from grazing. His idea of infrequent, short-duration grazing evolved into the multipaddock rotational grazing practiced by regenerative ranchers today.

Voisin's insight was to combine recognition that grass needs time after grazing to rebuild its foliage with the realization that cattle do best grazing fresh growth. The key to highly productive grazing lay not in how many animals one could stuff onto an acre but in how long cattle grazed a paddock and how much time passed before they returned to it. His cows produced more milk and stayed healthier when they grazed each paddock for just a day before moving on to the next. Voisin saw grazing animals as a tool to harvest the pasture's main crop—grass. A dairy farmer's real job was to grow forage; the cow's was to harvest and process it. He recognized three stages of grass growth: an early time of slow growth, a central spurt of rapid growth, and a final period culminating in seed production. Voisin thought the ideal system would allow plants to get through the first two phases before bringing in cows to graze at the start of the third phase.

His logic remains compellingly simple. When grass is shorter than several inches, the plant invests in the underground infrastructure it takes to support the next, rapid growth phase. In this second phase the plant pushes out fresh vegetative growth we now recognize as rich in omega-3s. In the third phase the plant forgoes new growth in favor of reproduction and building strong, relatively indigestible, cellulose-rich structural supports to hold up seed heads. Voisin's insight lay in seeing that maximizing the production of fresh grass meant trying to keep a pasture in the rapid growth of phase two, growing as fast as possible for as much of the year as feasible.

Cows allowed to graze for too long in a paddock didn't let plants get beyond phase-one growth. And it takes a while for a plant to rebuild the foliage and root system to support the accelerated growth of phase two. So moving cows frequently gives plants time to recover and keeps the cows grazing on fresh growth. Instead of letting them graze at will throughout one large paddock, frequently moving cattle between a lot of smaller paddocks defined a new approach that evolved into what we now call rotational grazing.

Voisin saw animals as biochemical photographs of the soil from which their food grew, and he thought that soil organic matter acted like a catalyst for soil life. Earthworms and microbes, he believed, needed to consume soil minerals and incorporate them into humus before plants could readily take them up and pass them along to grazing animals. He saw soil life as the key to maintaining fertility and keeping nutrients on the move, cycling from soil to plant to animal.

Voisin summarized his findings, theories, and experience along these lines in his 1959 book *Soil, Grass and Cancer*. He emphasized the importance of different mineral elements in the health of plants, animals, and people, tracing the effects of deficiencies through to botanical, zoological, and human health. He warned of a fundamental danger in simplifying the first step along this chain. Relying on a few chemical fertilizers ignored the need for other elements, like micronutrients essential for health.

In particular, Voisin thought overapplication of mineral fertilizers could throw off the balance of minerals in the soil. Fertilizers needed to be applied judiciously. Too much or too little of certain ones led to a range of health problems and conditions in ruminants. An imbalance of minerals affected the metabolism of cells, and a surplus or deficiency in one mineral could affect the availability or utilization of another.

For example, adding too much potassium to a pasture could reduce magnesium and calcium uptake, turning a surplus of one mineral into deficiency in others. While some went further in promoting the concept of an ideal balance between calcium, magnesium, and potassium as the key to maximizing plant growth, soil scientists today tend to discount the idea of a universally ideal balance for all crops and soils. Still, the actual mineral balance in a soil does matter, and Voisin's idea that micronutrient deficiencies affect the health of plants, livestock, and people remains as insightful today as it proved radical in his day.

Long before science understood metabolism and the catalytic influence of enzymes, people intuitively considered animals a reflection of the soil. Building on this traditional wisdom, Voisin considered trace minerals that moved from soil to plant to animal as catalysts for catalysts. He saw two in particular, copper and boron, as enabling biochemical reactions central to health. Lack of them, Voisin believed, cascaded disruptively through animal physiology.

He pointed to the availability of boron influencing the amount of the essential amino acid tryptophan in plants, the only source in the diets of animals and ultimately people. Low boron levels in the soil translated into insufficient levels of tryptophan in forages and the animals that ate them. Similarly, Voisin linked plants short on cysteine and methionine, two important sulfur-containing amino acids, to soils with low levels of sulfur. He also thought that too much nitrogen fertilizer would suppress sulfur availability and reduce levels of critical amino acids in livestock forage.

Yet unlike Sir Albert Howard, Voisin did not reject the use of chemical fertilizers. He was convinced that a favorable balance of minerals was critical for the health of plants and the cows that grazed them. While it did not help to blindly overapply the growth-promoting troika of N, P, and K, the key was finding the right balance for each soil. He called not for giving up nitrogen fertilizers but for using them more thoughtfully.

Voisin also recognized that conventional soil tests did not evaluate how much of different minerals a grazing animal could actually assimilate. Soil biology was as important as soil chemistry because it helped make vitamins and minerals available to plants and thereby to livestock and people. So he urged agronomists to consider the feeding behavior of cows. What scientists considered an ideal meal for cattle might be assessed differently by those doing the eating.

Arguing that the health of animals and people are linked to the mineral balance of the soil, Voisin offered numerous examples of cases where deficiency in one element or another translated into various maladies. Based on case studies in Europe and North America, he came to the radical conclusion that veterinary medicine needed to focus first on healing the soil so as not to have to heal the animal. Preventive medicine was the best solution for livestock health. And the key to that? Soil health.

Like his contemporaries we've already met, Voisin's insights landed well ahead of their time in recognizing how conventional practices that separated cows from grass disrupted the dance of grazing and regrowth that kept the land productive. In his view, housing cows in barracks rather than frequently moving them among small paddocks limited the productivity of both animal and pasture. Voisin also attributed the root cause of tuberculosis outbreaks among cattle to feeding them poor-quality forage in crowded feedlots. He saw these conditions compromising their health to the point that they succumbed to diseases at far higher rates than healthy animals.

Major illnesses and chronic diseases can take a big bite out of meat and milk production. In the best case they are a drag on yield. In the worst case they can cause the premature death of an animal. This leads to an obvious question: to what extent can animals choose a diet that helps them remain healthy?

Wild Health

Animals in the wild that become ill routinely choose foods with medicinal effects to make themselves feel better. Such behavior has a name—zoopharmacognosy. A mouthful of a word, its roots reveal its meaning: *zoo* (animal), *pharma* (remedy), and *gnosy* (knowing).

Through the ages healers and philosophers from ancient societies knew animals used the botanical world as a pharmacopoeia. Native Americans in the northern Great Plains, for example, refer to lovage, a plant in the carrot family, as "bear medicine." Upon emerging from hibernation, bears dig it up and eat the roots, apparently to get their gut moving again.

Jane Goodall's research along the shores of Lake Tanganyika in the early 1960s enthralled and shocked many when it revealed that chimpanzees could make tools and kill other animals for their supper. But they had another skill, too.

An assistant of Goodall's, Richard Wrangham, reported an odd event in his field notes one day. While following two male chimpanzees, Hugo and Figan, along a trail they regularly used, Hugo abruptly plunged off into a thorn-infested thicket. It was Wrangham's job to watch what the chimpanzees did. So he beat his way into the thicket and found Hugo plucking leaves from *Aspilia rudis*, an herbaceous plant in the daisy family. The chimp put a few leaves in his mouth, barely chewed them, and then swallowed. Wrangham had seen a lot of chimpanzees eating, and this was unusual. Most of the time a chimp ate fistful after fistful of leaves in a chewing marathon.

But Hugo knew exactly what he was doing. The surface of *A. rudis* leaves are like sandpaper. He wasn't looking to scrape up his tongue; he was looking to scour parasitic worms and nematodes out of his gut. Local Africans knew about the plant and used it as well. In the decades after Wrangham's discovery, chimpanzees, bonobos, and lowland gorillas from other parts of Africa were observed eating the leaves of almost 20 other species with similar scour power.

Wrangham later collaborated with a biochemist to investigate whether the phytochemicals of other *Aspilia* species might also be contributing to its antiparasitic effects. They discovered that even a minimally chewed leaf held enough of a particular phytochemical to kill about 80 percent of parasitic nematodes in a typical chimpanzee.

Can ruminants do things like apes? It turns out they can. For decades, Fred Provenza has explored how they deploy their innate behavior and inherent biology—their body wisdom—to meet their needs for balanced nutrition while reaping a range of health benefits from phytochemicals and other compounds in plants. From the deserts of the American Southwest to emerald pastures in New Zealand, from traditional shepherding landscapes to the most miserable confined indoor feeding operations, he and his colleagues have documented how ruminants acquire, use, and refine their body wisdom—or not. This insightful research has reshaped assumptions and conventional thinking about the capabilities of ruminants when they choose a diet of living plants growing in healthy soil.

Three Pillars

Three things are vital for body wisdom. Learning is one such pillar, much of which takes the form of interactions within a herd, flock, or pack—and especially between moms and their young. Flavor is the second pillar, driving a ruminant's food preferences through different types of feedback. The third pillar relates to place, about whether or not the environments in which animals live offer opportunities for them to engage the first two pillars. Connections among the pillars occur at multiple levels from cognition to cellular communication. Extensive research shows that when given a chance to use their body wisdom, herbivores excel at deciphering and navigating the richness of the botanical world to find nourishing food and medicine. And this in turn translates into an animal's health and welfare, as well as nutrient-dense meat and dairy.

Ruminants began schooling Provenza decades ago when he worked on a ranch during summers in the Rocky Mountains near where he'd grown up in south-central Colorado. After earning an undergraduate degree in wildlife biology, he decided to pursue a graduate degree. He envisioned hiking through lush alpine meadows filled with wildflowers to reach the high-elevation home of mountain goats white as snow.

Like many graduate students, Provenza found things didn't go quite like he planned. As it turned out, he soaked in sweat at Cactus Flat, a desert study site in the southwestern corner of Utah, traipsing after domesticated goats while dodging spines, stickers, and thorns. Provenza hoped to learn how the browsing behavior of goats might be managed to improve forage for wildlife and domesticated ruminants.

But before he got the goats, he had to build a set of pens for them. He spent most of the first summer at Cactus Flat in torrid heat digging one four-foot-deep hole after another in desert hardpan soils to set posts for two miles of fence. Fortunately, his wife, Sue, agreed to pitch in. Should you ever have the pleasure of meeting Fred and Sue, you'll see neither are complainers, but that summer at Cactus Flat, they both agreed, was hell. Still, thankless or not, a task is a task, and by fall they completed building six paired pastures to house the goats.

Provenza came to appreciate small details of the landscape, especially the plants. One of the most common was blackbrush (*Coleogyne ramosissima*), a member of the rose family. Its small, thick, sage-green leaves sit snugly along spiny stems, and the whole shrub tops out at about three feet. Goats are known for eating plants that look unappealing to people. Blackbrush fit the bill.

One of the pastures contained particularly lush blackbrush shrubs growing around a once-stately juniper tree recently killed by lightning. Provenza figured the goats would feast on this primo forage, as close to caprine heaven as he imagined they could find in a desert.

When 90 white Angora goats that Fred had leased from the Navajo Nation showed up, he and Sue spent the better part of a week getting them settled in their pastures. Sure enough, the goats in the pasture with the dead juniper noticed the juicy blackbrush, running their noses up and down twigs and sniffing the foliage of individual shrubs. Provenza waited for them to start their feast. They never did, although one solitary oddball did take a couple bites but then ambled off. Provenza, stupefied as to why the goats shunned such prime blackbrush, was haunted by a mentor's words: "You should never study an animal that's smarter than you are, and *you* shouldn't be studying goats."*

The goats were not done surprising him for they soon found nourishment elsewhere thanks to another Cactus Flat inhabitant endowed with inquisitive black eyes, oversized ears, and a long furry tail with a tuft at the end. Desert woodrats are masterful builders, using dead sticks and twigs to construct elaborate homes that can get as big as a Volkswagen beetle. The cleverness of these diminutive rodents impressed Provenza, especially their ingenuity in repurposing the bark of juniper trees as siding for their homes.

The goats considered woodrat houses self-serve pop-up restaurants. Provenza watched a goat tear through the juniper siding and make an opening big enough to shove its entire head through. Everything was fair game, from caches of fresh vegetation, seeds, and fruits to the long-dead sticks and twigs the woodrats used as their building materials.

* Provenza, F. 2018. *Nourishment: What Animals Can Teach Us about Rediscovering Our Nutritional Wisdom*. Chelsea Green Publishing, White River Junction, VT, p. 16.

Although productive, the goats' foraging expeditions were messy affairs. Eating a woodrat abode turned their muzzles a snappy, two-tone color scheme that looked as if painted on—browns and blacks at the front crisply transitioning to their natural white near the back of their jawline. This new look made it easy for Provenza to tell which goats learned how to eat woodrat houses.

But one thing didn't make sense. Most of what the goats ate wasn't even food; it was the dead, woody material composing woodrat houses. What possible nutritional value did it hold? The answer, as it turned out, was nitrogen. The goats were onto it way before Provenza. Woodrat urine permeated the building material and contents of their abode. After the water in the rat pee evaporated, nitrogen-rich urea remained. And just as for plants, nitrogen is a must-have for animals. In the goats, it would support rumen microorganisms, go toward building muscle, and patch up damaged DNA.

Still, solving one goat riddle led to another. Provenza knew that the blackbrush also contained nitrogen, so why weren't the goats eating the plants they were supposed to? All the lush, young growth, Provenza would later document, harbored high levels of noxious tannins that kept the goats at bay. A lot of plants do this to protect their new growth from herbivores. The goats preferred devouring a desert woodrat house soaked in edible nitrogen to eating living plants laden with phytochemicals that could make them sick if they ate too much.

Provenza's experience with the goats at Cactus Flat set him on a career-long journey. He went on to become a professor at Utah State University, where he focused on the fundamental question of how animals know what to eat and make food choices to meet their nutritional and medicinal needs. It turns out that the body wisdom of ruminants can teach us about animal husbandry and agricultural practices—as well as how to nourish our own bodies.

Flavor Feeds Back

This view was not at center stage when Provenza started out as a graduate student. His professors held that it was risky business to raise domestic ruminants in range and pasture settings. Agronomists and animal scientists saw forage plants as reservoirs of toxic phytochemicals that could interfere with a ruminant getting a nutritionally adequate and balanced diet. At first, Provenza accepted this idea. Yet the more he observed and documented animals making food choices and developing eating patterns, the sharper the complex truths of ruminant life came into focus. Their success or failure to thrive grew out of experiences reflecting the interplay of their biology, behavior, and interactions with the place they lived. Plants, soils, water, and other animals were friend and foe, challenger and ally. When ruminants get to choose a diet from a diversity of living plants, their body wisdom juggles the perils and benefits of the botanical world as it has done for millennia.

Tall fescue, a common forage grass, illustrates this point. It is a good source of protein for cows, as well as a source of the complex carbohydrates (fiber) that keep rumen dwellers productive and the rumen functioning normally. But toxic alkaloids that come packaged along with the protein and fiber are often all that stands between a tolerable level of herbivory and grasses getting grazed into oblivion.*

A domesticated herbivore can eat some alkaloids, but if it ingests too much, blood vessels constrict in peripheral parts of the body, starting with the ears, tail, and hooves. This effect explains colloquial names for alkaloid poisoning like "ryegrass staggers" and "drunken horse grass." If the poisoning goes on long enough, constricted blood vessels lead to tissue death and ultimately gangrene, which can prove fatal.

---

* Interestingly, many grasses, including tall fescue, don't make their own herbivore-deterring compounds. Instead they provide bacterial and fungal symbionts with exudates and receive alkaloids in return.

So what's a ruminant to do? They need the nutrients in grasses, but those nutrients come intertwined with troublesome alkaloids. Provenza and one of his graduate students set up a study to find out how cattle learn to avoid overdosing on alkaloids. It turned out to be pretty simple: their bovine study subjects chose to eat plants that contain alkaloid-neutralizing phytochemicals before eating plants with alkaloids.

For two 12-day periods they let cattle graze on alkaloid-rich grasses (reed canary grass and tall fescue) first thing each morning. Immediately afterward, the cattle were moved so they could graze two other common forage plants—alfalfa and birdsfoot trefoil. As the twelfth day of the trial approached, the amount of alkaloids in the cattle had built up to a high-enough level that they ate less and less of the grasses, especially the tall fescue. Body wisdom was at work. The cattle associated the taste of grass with feeling unwell: the flavor-feedback and learning pillars together led cattle to eat less of the fescue.

During the second 12-day period, first thing every morning the cattle grazed alfalfa and birdsfoot trefoil before they grazed on the grasses. This did the trick. Alfalfa contains saponins while birdsfoot trefoil contains tannins. Both of these phytochemicals bind to alkaloids in the ruminant gut, which keeps levels low in the bloodstream. So long as the cattle first ate the plants containing the antidote—the saponins in the alfalfa and the tannins in the birdsfoot trefoil—it neutralized the toxic alkaloids in the grasses. This had another effect ranchers are naturally interested in. More cow. Animals that don't feel bad or get sick keep eating—and growing.

A similar study carried out with sheep produced the same result. Eating alfalfa or birdsfoot trefoil before tall fescue or reed canary grass improved their digestion and led the sheep to eat more grass. Ruminants store such experiences in their memory so that the next time they encounter a food, they can gauge how much of it to eat, what else to eat it with, and in what sequence to avoid becoming nauseated or sick.

What these studies (and many more like them discussed in Provenza's book *Nourishment*) reveal, is that when ruminants eat a diet rich in a diversity of living plants, the process feeds on itself, so to speak. Animals raised in such settings constantly adjust the mix of forages to meet their unique needs for nutrients, minerals, vitamins, and other beneficial compounds while not overeating plant toxins.

Some years after his time at Cactus Flat, Provenza decided he'd do a feeding experiment to nail down which phytochemicals imbued blackbrush with goat-deterrent powers. With the help of a colleague, he extracted and purified candidate compounds from young blackbrush twigs and incorporated them into separate batches of food pellets. Over the course of several months, he fed one batch at a time to a group of goats. Every time, the goats ate all the pellets.

Provenza came down to the final batch, the one containing tannins. He figured this must be the phytochemical that kept the goats away. Alas, they wolfed down the tannin-rich pellets! Provenza watched in disbelief and frustration. He had exhausted precious grant funds on the blackbrush extract work and expended considerable time and effort on the feeding experiment. Was it all a big, fat nothing?

There were only enough tannin-rich pellets in the batch for one more feeding. So, the next morning he offered the last of the pellets. Plink, plink, plink into the food trough they went. Fred watched the goats, and they stared back at him, thoroughly disinterested in eating. What they chowed down on the previous day now had all the appeal of wet cardboard. Once again, the goats led. But where, he wondered, were they taking him?

Provenza realized the goats were showing how the flavor pillar of body wisdom worked as a feedback system. Like the cattle that had their fill of alkaloids from eating the tall fescue, goats and other animals readily associate a particular flavor with how they feel after a meal. Should malaise or some other negative outcome follow on the heels of eating a particular plant, a ruminant stops eating it for a time.

That's why the goats showed no interest in the second meal of pellets with tannins. Then when negative feelings subside, a ruminant will return to eating the plant. This association works the other way as well. When a positive feeling follows a foraging session, an animal seeks out the same plants again in the future. A dynamic, ever-changing dance with the botanical world is as embedded in the ruminant lifestyle as placid cud chewing.

As with most phytochemicals, tannins have multiple effects on animals, including beneficial ones like parasite control. Domesticated ruminants use the same basic strategy of the great apes: knock back parasite populations so that the effects are minor rather than life-threatening. Sheep employ it to deal with barber pole worm (*Haemonchus contortus*), a notorious nematode that plagues flocks around the world. The common name of this parasite comes from the circular pattern of red and white striping that fills their tiny translucent bodies as hordes of them latch on to the gut wall of a ruminant, pirating meal after meal of nutritious blood. This situation can quickly progress to anemia and, if left untreated, can kill an animal.

Deworming drugs are the typical treatment, but in many places they no longer work against barber pole worm. This is the situation for many other parasites too, with over half the farms in the United States, Australia, and Brazil facing drug-resistant parasites.

Fortunately, tannin-rich forages help on this front, enhancing an animal's immune response to parasites, disrupting egg and larval development, and dislodging parasites from the gut wall. Tannins also have antimicrobial, antiviral, and anti-inflammatory properties. Their effects can be medicinal or hazardous depending on a number of factors, including the type and amount of other phytochemicals already present in an animal, what they eat before and after the tannins, and their overall health. Interestingly, our current understanding of the science on tannins reinforces the "dose-makes-the-poison" adage cred-

ited to the peripatetic sixteenth-century Swiss physician and alchemist Paracelsus.

A major advantage of using tannins to control parasites is that it doesn't wreak havoc on soil life. Many pharmaceutical dewormers remain lethal in the feces of drug-treated animals and prove disastrous for dung beetles in particular. Controlling parasites through allowing animals to eat diverse pasture forages that include tannin-rich species offers a way out of this problem.

Average Problem

Body wisdom, with its compass-like qualities geared to an animal's unique biology and metabolism, proves a challenge to those charged with formulating TMRs. These folks run right into one of the most common, universal, and vexing problems in product design. Bovine or human, every body is different. Consider, for example, the driver's seat of a car. Design one for an average-sized person, and it won't work well for everyone. After over a century of making cars the best we've come up with is hyperadjustable seats—assuming you can figure out how they work.

Formulating TMRs for animals in confined feeding environments remains similarly complex. Individual animals have different metabolic rates and unique biochemistry, all of which can fluctuate with life history stage, season, health status, and other factors. The reality is that a product designed for an average person or animal (or plant) does not work well in a world full of unique individuals with their variable sizes, metabolism, genes, and microbiomes. Think about it: does everyone in your family always eat the same amount of the same thing for every meal?

Diversity persists in domesticated animals and plants despite our efforts to standardize them. All that diversity, however, is actually a good thing. It builds resilience to challenges like pathogens and enables some animals to eat more of certain plants that others can't tolerate well. Variability may not always work in favor of an individual animal, but over time it enhances the resilience of a population. Body wisdom is part of this adaptive process. Letting it work is a sound way to support the health of farm animals.

With this perspective in mind, what do you suppose a ruminant would do if given the choice to eat the individual ingredients that go into making a TMR? Would it choose a nutritionally adequate diet or just pig out on its favorite ingredient? Researchers carried out such a study on calves. One group had free access to TMR ingredients—alfalfa, corn silage, rolled barley, and rolled corn—laid out in separate bins. The other group received these same four ingredients ground and mixed into TMRs.

Several things were notable in the group eating the individual ingredients from separate bins. From one day to the next, no calf ever chose the same combination or the same amounts from among the four ingredients. Today, rolled barley may be a calf's favorite with just a bit of the other three ingredients. Tomorrow, the same calf might eat alfalfa and corn and skip the barley. The calves' choices also never matched the same proportions of ingredients as found in the TMR. Some days a calf might eat more protein, other days more carbohydrates. In the end both groups gained weight at the same rate and remained in good health through the several-month trial. The free-choice diet, however, cost the farmer less. Calves in this group happened to eat less of the most expensive ingredient.

Convenient and standardized, TMRs form the dietary backbone of indoor feeding operations under the assumption that standardization is efficient and thus cost-effective. But like people, there is no average animal, only individuals and their unique biochemistry and metabolism. So if one animal needs more of a vitamin or a particular amino acid or mineral than the standard amount in a TMR, it has to overeat to get enough of the one nutrient. Though little recognized, this reality translates into higher feed costs for producers.

Formulating feeds based on an average animal can lead to undereaters. These individuals have less of a need for the amount of a particular nutrient in TMRs. To avoid feeling unwell or becoming sick from eating too much of it, they eat less of their TMR. Depending on the degree of undereating, an animal can become underweight or more prone to other health problems.

Palatants can further compound the problems of animals that eat too much or too little on a TMR diet. These appetite stimulants short-circuit an animal's body wisdom using flavors to alter normal satiety cues. Animals that undereat TMRs to avoid a particular ingredient, or ingredients, are likely to feel even worse if they eat more of what is making them unwell. And animals already overeating TMRs to get enough of a single nutrient run the risk of becoming sick from ingesting too much of the other nutrients. In the end, for all their perceived advantages TMRs are also a recipe for extra feed and veterinary care costs.

Lest you think ruminants are the only ones eating palatants, consider the heavily processed food products people eat. Most consumers would object to finding "palatant" on a list of ingredients. So more appetizing terms are used, like "natural flavors" or simply "flavorings." Added sugars, fats, and salt are more recognizable and obvious palatants, with some proving particularly problematic. Artificial trans fats, for example, were initially thought safe but later banned from foods. Whatever they get called, flavorings added to processed foods to get us to eat more affect our food choices and behavior as surely as they do those of cows living on a diet of TMRs.

Re-envisioning how we raise animals begins with a pretty simple observation: today's norm is not normal. What if instead of trampling on a ruminant's body wisdom, we raised them in ways other than confining them to feedlots on grain- and concentrate-intensive diets laced with palatants? For nearly all of human history the keepers of domesticated herds guided their animals to pastures and foraging spots, for the most part letting their animals do the picking, choosing, and sequencing of what to eat. Still, people had to pay attention to linkages between plant health and soil quality. This, after all, was the key to the health of the herd and thus their own family and community.

This brings us to the third pillar of body wisdom—place—and the degree to which it allows animals to express and use the first two pillars. Given the right environment and ample opportunities, ruminants can use the flavor-feedback loop to learn about and select plant foods they need to maintain good health.

Animals of course are not moving through a pasture or landscape thinking about alkaloids, tannins, or any particular phytochemical. They gradually learn about a place and the plants that grow there through experiences over their lifetime. This aspect of body wisdom explains a number of things, including why cattle drives from the plains of Texas to the stockyard industry emerging in the Midwest in the late 1800s often ended with so few animals reaching the destination. Cattle had little to no familiarity with the plant communities along the route and often didn't get enough to eat. They also lacked knowledge of watering spots, natural hazards, and places to shelter. As a consequence, many perished along the way. This situation is equally problematic for populations of wild ruminants and other animals. Lack of familiarity with food and water sources, terrain, and so on is a major problem for reintroduction efforts.

The importance of place also figures into why, should you decide to buy a ranch, it typically comes with the resident cattle. Bringing in animals naive to a place usually doesn't work out well. They don't know their way around the landscape. Animals born and raised in a landscape know how to put their meals together based on the seasonal growth and availability of the local plant communities. Acquiring such place-based knowledge is a lifelong process and passes from generation to generation.

Lack of food choices and learning opportunities that build body wisdom are just the beginning of problems for calves routed to feedlots. Upon arrival, the knowledge they learned from their moms and other herd members in a pasture or range environment unravels. There is no picking or choosing what to eat. There is only one thing—TMRs—and one place to get it—the feed trough.

Since Voisin's time, agronomists have been choosing the plants they think ruminants should eat. Once animal agriculture transitioned from pastures and rangelands to feedlots we found ourselves trying to replace a ruminant's body wisdom with our own thinking. We applied human nutrition concepts to animal husbandry, formulating protein, fats, and carbs, along with selected vitamins and minerals into TMRs and other concoctions.

In practice this approach skips over a key feature of the ruminant diet—phytochemicals and how the pillars of body wisdom, especially at a cellular level, manage the dose-makes-the-poison dynamic. A diet of TMRs guts body wisdom, in part because it delivers relatively few phytochemicals compared to the levels in a diet of diverse living forages. Evidence from studies of pigs, poultry, and cattle indicate that plant polyphenols help reduce local and systemic inflammation, much like they do in our bodies. All in all, grazing phytochemically rich forages generally enhances cattle health compared to consuming feedlot rations. Grasses are good of course, but adding forbs and shrubs is even better. They add more terpenes, polyphenols, tannins, and carotenoids, fully stocking a ruminant's on-board pharmacy.

Long-Lived Diversity

No matter how much our eyes may tell us that if we've seen one cow, sheep, or goat, we've seen them all, such thinking belies one of the most fundamental tenets of biology. This takes us back to where things started—with Fred Provenza at Cactus Flat. By the end of cycling all 90 goats through the 6 pens, only a small number had learned the trick of eating a woodrat house to get at the nitrogen it held. These goats were outliers, the individuals of a group that don't behave as you expected. Far from average, they are the different ones. And it is from these differences that adaptation and resilience arise.

Diversity runs deep through the history of life. It provides the genomic variety at the heart of evolution. Rather than trying to crush this biological reality into uniformity, it would behoove us to accept and work with it. Even in domestic farm animals it is variety and diversity that imbue the individual members of a herd or flock with the ability to develop, express, and refine their body wisdom. This is also why nearly every dietary or feeding study, as well as any drug study that has ever been done on animals or people, all find outliers. No two animals or people are truly the same, not even identical twins. Every body is biochemically distinct, which makes nutrition and biomedical research maddeningly complex and inconsistent.

While others have given the problems of feedlots the more in-depth treatment they deserve, it is worth pointing out here that such settings also interfere with an animal's fundamental freedoms. Moving cows from their homes in familiar grazing settings to the unfamiliar and stressful environment of a feedlot violates their freedom from fear and distress. A diet known to lead to chronic ailments and premature death violates their freedom to live free from undue pain, injury, or disease. And confined animals fed such a diet experience stress and nausea, violating their freedom from discomfort.

In the end, beyond love and respect for animals there is another compelling reason to raise livestock in ways that provide the basic freedoms that engage their body wisdom to support their health. It translates into healthier foods—meat, eggs, and milk richer in phytochemicals and beneficial fats. What's good for them is good for the land—and us.

PEOPLE

FLAVOR OF HEALTH

You've got to take the bitter with the sweet.

—CAROLE KING

We all love a tasty meal. But there's more going on than simple pleasure. It turns out our own bodies possess some pretty impressive biological tools for detecting what is in the things we eat and drink. And just as for ruminants, flavors inform our body wisdom.

Think about the time long before grocery stores, pubs, or restaurants. What you ate came down to what you, your family, or your tribal group could forage and hunt. There are obvious advantages to liking foods that power your body, hasten healing, and help prevent disease. With the help of body wisdom, you'd remember those foods, where you found them, and how certain ones eaten together delivered benefits greater than when eaten separately.

Flavor and taste have always been means to an end. On the surface they were how we identified and remembered sources of nourishing food and avoided toxins. But on a deeper level, flavor and taste guide us to health-imbuing nutrients and compounds. Just like that of our herds and flocks, our body wisdom developed a nutritional dossier through repeated exposure to different flavors. The agents that collect and help interpret all this intelligence are special cells tucked deep inside the nose at the doorstep of the brain along with taste receptors in the mouth.

The tongue readily detects compounds dissolved in saliva that we recognize as one of five tastes—sour, sweet, salty, bitter, and umami. Sour signals acidity, like that of vinegar or unripe fruits, as well as certain phytochemicals like those in citrus. Sweet means carbs and thus energy, while salty tastes help our bodies regulate electrolytes.

Although research on its origins began over a century ago, umami, the savory taste, is a latecomer to the gustatory roster. Japanese chemist Kikunae Ikeda found that none of the then standard categories of taste captured the dominant taste in dashi, a liquid stock flavored with kombu (seaweed) and thinly shaven bits of smoked fish. He began analyzing what was in these two ingredients and by 1908 identified that glutamate, a salt formed from the amino acid glutamic acid, played a leading role. Almost a century later, in the early 2000s, the discovery of umami taste receptors completed the story. Umami-rich foods include protein sources like meats and cheeses, although certain vegetables bring umami too, notably mushrooms and tomatoes. Combine either of these vegetables with meat or fish, as is done in many cuisines, and the ensemble delivers rich and satisfying flavors.

Some scientists think fats also warrant designation as an official taste given the presence of cell receptors for them in the mouth. This maybe-a-taste already has a proposed name too: "oleogustus," Latin for the taste of oil.

Bitter tastes come from phytochemicals, with a twist linked to one's genes. Some people are highly sensitive and can detect bitter tastes in minute amounts. Others are far less sensitive and find bitter substances, even at relatively high levels, perfectly palatable.*

While we perceive taste through receptors on our tongue and other places in the mouth, flavors flow from a complex merger of taste and smell. This union is possible thanks to mammalian anatomy. Our mouth and nose are not walled off from one another like separate rooms. Receptors in the nose pick up on fleeting and unique arrays of volatilized compounds released as we chew food, slurp liquids, or open the oven door to take in the aromas released as food braises, bakes, or broils.

These airborne molecules waft upward from the moist cauldron of the mouth, winding through cavernous labyrinthine terrain until they reach olfactory receptors located about as far up and back in the nose as they can go. At this biological end-of-the road sits the olfactory bulb, a small part of the brain.

The direct connection between olfactory receptors and the brain is no accident. Smells, aromas, and scents carried through the air bring detailed information about food and the environment straight to our body's cognitive headquarters for evaluation. The vast breadth of our sense of smell is hard to fathom. We have about 400 different olfactory receptors capable of detecting a trillion distinct aromas. To put the human senses of taste and smell in perspective, consider tastes akin to a single colored tile—and flavors more like an intricate mosaic.

There are other important aspects of taste and flavor, too. We rely on physical properties of food like temperature, appearance, and texture to feed our senses, while social and aesthetic conditions also play a part. Something you normally consider ho-hum tasting may go down much better at a celebratory event like a wedding or a social gathering

---

* Those with the gene variant that confers the highest sensitivity to bitter substances are "supertasters" and comprise about a quarter of the population. At the opposite end of the spectrum are "nontasters" with the least sensitivity. They also represent about a quarter of the population. "Tasters" are in between and account for about half of humanity.

in the company of friends over a glass of good wine. Conversely, even a food you normally love may hold no appeal if you were locked in jail.

The way food looks also affects what we expect in terms of taste and flavor. Take tomatoes. We've all been disappointed biting into one that looked great but lacked flavor. But a sense of smell and taste that quickly detects lackluster foods serves a higher purpose.

Bland-tasting fruits and vegetables, it appears, contain fewer health-protective compounds than those suffused with flavor. At least that's according to a 2006 study published in *Science* that analyzed hundreds of volatiles in tomatoes. It turned out that the volatiles linked to the most satisfying flavors are rooted in compounds that directly benefit human health, like carotenoids—phytochemicals essential for sound vision and eye health. The essential fatty acids, LA and ALA, contributed to the best-tasting tomatoes too. In addition, three amino acids (phenylalanine, leucine, and isoleucine) also underpinned mouth-watering flavors. These particular amino acids are among those that our bodies cannot make. The only naturally occurring source of them is the foods in one's diet. This is but one example of how taste and flavor feed body wisdom, guiding us toward health-imbuing foods with no conscious effort.

The flavor profiles people find appealing in other fruits, vegetables, herbs, and spices also spring from the same broad categories—phytochemicals, essential fats, and amino acids. Those rubbery, bland tomatoes and tasteless veggies really are less healthy than truly delicious ones. In the end, gussied-up fruits and vegetables may delight the eye, but they shortchange health.

Learning about flavors early in life builds experience we rely on ever after. Flavor drives why we eat and drink something over and over or never again. It lies at the heart of our strongest preferences, like a specific version of Thanksgiving stuffing, a favorite spaghetti sauce, or a hearty winter stew. But flavor can stir controversy too. We have all likely dished out or received vociferous complaints about an ingredient substitution or makeover for a favorite family dish.

Flavor feedback in humans starts in the womb and builds throughout life. A study on pregnant women showed that consuming carrot juice led to a preference for carrot flavor among their babies. In the last trimester of pregnancy or while lactating, a group of women were assigned to drink a cup of carrot juice four days a week, while another group only drank water. Once weaned, all the babies received cereal with carrot flavor. Of course the babies could not say what they thought about the carrot-flavored cereal, but the looks on their faces told the story. Those whose mothers drank carrot juice made positive facial expressions, like the start of a smile or a full-on grin, while the faces of the babies whose mothers drank only water did not. We're not the only mammal whose early-in-life flavor experiences stick with us. The same flavor feedback occurs in ruminants. What moms eat strongly influences food preferences and related behaviors in their offspring.

Of all the tastes in the human diet, those of bitter substances best illustrate the multifaceted capabilities of our taste receptors. Conventional thinking about bitter tastes maintains that their purpose is to warn us about toxic phytochemicals and other harmful compounds in food. That view, however, captures only part of the story behind our long-running relationship with plant foods.

Take the case of glucosinolates. These phytochemicals contribute to the bitterness of broccoli, kale, and other plants in the mustard family. In very high amounts, they can disrupt thyroid hormone production, although you would have to eat a nauseating amount of broccoli to produce such an effect. On the positive side, a substantial body of evidence shows that regular dietary intake of glucosinolates in vegetables spurs cellular activity that helps prevent the onset of chronic diseases like cancer and heart disease.

So how is it that we navigate the duality of bitter tastes? Through a combination of cognitive power and linked-at-the-hip interactions between bitter taste receptors and subconscious controls on our physiology and metabolism. Much like the pillars of body wisdom at work in ruminants, our cells and organs respond to internal cues ranging from hunger and satiety hormones to information stored as memories about positive food-associated experiences. Working in this way, body wisdom guides us to take in foods with phytochemicals in amounts that benefit us.

The dual-purpose nature of body wisdom becomes even clearer in the context of an odd, little-known fact. Bitter and sweet taste receptors found in the mouth also pepper nervous, endocrine, and immune cells in our gut. Of course, they interpret taste and flavor differently than what you consciously sense. They are more attuned to things like levels of vitamin C or a specific phytochemical. At the same time, bitter taste receptors also scan for harmful levels of toxins.

In due course, and depending on what you ate, receptors in the gut pass along information embedded in a meal to cells and tissues that set other processes in motion, like activating endocrine cells in the gut that sync the release of insulin with food intake. This way sugar gets shuttled out of our blood and into cells at the right pace. Stimulating a taste receptor on particular gut-associated immune cells helps tee up inflammation-quelling action. Taste receptors in the gut even play a role in the muscular contractions that keep intestinal contents moving along.

These are but a few examples of how the gut serves as much more than a place of digestion. Cascading from diet to taste receptor to internal dossier, flavors arising from combinations of fats, phytochemicals, and amino acids initiate an intricate flow of information and feedbacks that ripple through every system of our body, driving behaviors that underlie food choices and ultimately health.

Along with bitter and sweet taste receptors, olfactory receptors also occur in internal organs and tissues far beyond the mouth and gut, including bone marrow and our hearts, testes (if you have them), skin, bladder, and the brain, as well as the immune and respiratory systems. According to one study, when exposed to the aroma of cultured butter, particular immune cells move from one location to another in a manner similar to how they mobilize at the site of an infection. And while it might seem weird, our skin has a sense of smell, so to speak. Research on olfactory receptors in human skin-cell cultures showed that a synthetic version of sandalwood aroma increased their growth rate in ways consistent with how flesh wounds heal. Such research expands Hippocrates's well-known adage "let food be thy medicine," showing that what lands on our plates guides the behavior of cells to help keep us humming along as we should.

Bitter receptors are abundant in airway tissues and offer a particularly rich picture of how bitter substances in foods influence human health. In cell culture experiments, flavones, common polyphenols found in many fruits and vegetables, increased the efficacy of antimicrobial compounds that human airway cells secrete to clear common pathogens. And in randomized clinical trials, a commercially available herbal extract rich in flavonoids proved effective as an adjunct therapy to use with antibiotics and decongestants for treating inflamed sinus and nasal cavities (chronic rhinosinusitis). With about 20 percent of all antibiotic prescriptions written for this condition, activating bitter taste receptors holds potential to help reduce antibiotic use and slow the rise of drug-resistant pathogens.

Interestingly, bitter taste receptors in airway tissue also respond to pathogens that ride into our bodies as we breathe, triggering a cough, sneeze, or swallow to expel them. But if a pathogen manages to stick around, bitter taste receptors can boot them out another way, through changing the behavior of cilia, tiny hair-like structures protruding from cells lining the airways. Day after day cilia sweep excess mucus and particulate matter out of the respiratory tract. But when a pathogen shows up and activates the bitter receptors of a ciliated cell, it causes the cell to release nitric oxide, which transforms its cilia into herculean sweepers. Nitric oxide, as you may recall, also has corrosive powers. Super-sweeping cilia spread it across airway tissues the way a fan moves air across a room, killing or disabling pathogens caught up in the toxic wave.

A study during the COVID-19 pandemic found that nitric oxide helped suppress viral replication. This finding suggests that tapping into bitter receptors in the respiratory system to trigger cilia super-sweepers, and thus pathogen-killing nitric oxide, offers a new approach to developing treatments for coronavirus infections in general.

Along these lines, a 2020 review of animal, human, and cell culture studies presented evidence that some phytochemicals are effective against serious infections arising from COVID-19 as well as other coronaviruses. Several common dietary phytochemicals appeared especially effective, including kaempferol, resveratrol, and quercetin. These phytochemicals helped knock back coronaviruses through activating anti-inflammatory processes and protecting the ciliary cells that keep airway tissues free and clear of pathogens.

The importance of bitter taste receptors in respiratory health raises a fundamental question. Do agricultural practices that decrease the amount of polyphenols and other phytochemicals in crops deprive our bitter taste receptors of substances needed to activate them?

Consider that humans have at least 25 different types of bitter taste receptors. Not so for the other four basic tastes, each of which has only several types. Why would we need so many different types of bitter taste receptors? Look to the botanical world. There are several hundred thousand plant species, each capable of making hundreds to thousands of phytochemicals. And though we eat far fewer wild plant foods than we once did, our bodies hung on to the biological tools for detecting and interpreting the immense array of phytochemical diversity in plants. In contrast, sweet is pretty unidimensional, signaling the presence of energy-rich foods.

There is also a highly personal angle to the role of bitter taste receptors. *Pseudomonas aeruginosa*, a well-known bacterial pathogen, can wreak havoc in the lungs, gut, outer ear, and urinary tract. Those who possess the gene variant that confers a high sensitivity to bitter tastes are lucky. Their super-sweeping cilia activate more quickly, clearing a *P. aeruginosa* infection sooner. On the other end are those with the least sensitivity to bitter tastes. They tend to get more severe cases of chronic sinus inflammation and are more susceptible to certain types of respiratory infections. It's not a stretch to think that declining levels of phytochemicals in crops may disproportionately affect those of us who have the least sensitive bitter taste receptors.

While such findings are thought provoking for human health, think too about the impacts of reducing the diversity of living plants in an herbivore's diet. Has this undercut the connection between bitter substances and their immune response? Though speculative, this may help explain the higher rates of bovine respiratory disease in feedlot settings.

Differences in response to bitter substances appear to have been at play during the COVID-19 pandemic. In a study of almost 2,000 subjects in their 40s, supertasters were found less likely to become infected, and if infected, they exhibited symptoms for fewer days and were less likely to be hospitalized.

Perhaps the most compelling evidence for the importance of interactions between bitter substances and human health is that pharmaceutical companies use bitter taste receptors and the hundreds of different types of olfactory receptors distributed throughout the body as targets for drugs. Such efforts look promising for those with the gene variant that makes them more susceptibile to airway disease. But let's not overlook what our bodies can do with the right diet given the long-running interactions between receptors and bitter substances in the foods we eat, particularly those found in whole vegetables, fruits, herbs, spices, and other botanicals. After all, flavor feedback in human cuisines and cultures is part of what led to the unique food and spice combinations of traditional diets.

Like the receptors that detect bitter tastes, those that detect fats in our diet create an equally impressive cascade of physiological effects. Whether or not fats ever make it onto the official roster of tastes, research shows that people have specific receptors that detect short-, medium-, and long-chain fatty acids as unique and distinct from one another. These receptors occur in taste buds as well as every major organ system, and studies on rodents and human cell lines reveal that activated fat receptors regulate important aspects of immunity, glucose metabolism, and hormone production.

Certain fats can also trigger olfactory receptors found well beyond the nose that influence health. A particularly telling example comes from a study that investigated the effects of short-chain fatty acids on kidney function. These fats activate olfactory receptors located in the kidney that regulate the release of renin, a hormone that plays a major role in regulating blood pressure.

The origin of short-chain fatty acids makes this bit of research particularly illuminating with respect to diet. Human gut bacteria produce them through fermenting the fiber in plant foods that we are unable to digest. The short-chain fatty acids then pass through the gut wall into the bloodstream. This is all part of the mounting evidence that along with what our food ate, we are also a product of what our gut microbiota eat. Research on other metabolites that gut microbiota make when a person's diet contains an abundance of fiber and phytochemical-rich plant foods suggests that they help reduce the risk of psychiatric and neurodegenerative diseases. Like the four-legged herds long a part of human cultures, what our internal herds eat also plays a pivotal role in our health.

Broken Compass

Body wisdom served humans remarkably well for nearly all of our quarter-million-year existence. It enabled our ancestors to live on a diet sourced from local plants and animals in the hottest, coldest, driest, and wettest places on the planet. If you consider DNA as the code for building life, think of body wisdom and the pillars that support it as the code for taking care of life, for maintaining health. The beauty and power of a code as time-tested as body wisdom lies in its high level of predictability and responsiveness. As useful as these qualities may be, though, they also render body wisdom vulnerable. Once its code gets cracked, it's easy to manipulate.

Our taste and smell receptors get us salivating over winning combinations of compounds that make for the best-tasting tomatoes. But when growing or breeding practices reduce the flavor of fruits and vegetables, there is little for body wisdom to work with or evaluate. This diminishes the nutritional dossier on which body wisdom relies, and so the flavor-feedback loop falters. And though there is plenty of disagreement as to the best source of the best flavors, some common ground unites us all. Flavorists, food marketers, manufacturers, and of course consumers all despise flavorless food. So what's the fix? Add flavors. Unfortunately, our body wisdom also falls for them.

It is no accident that people like the engineered flavors of ultra-processed food products. The very best flavorists in the food industry make us clamor for their creations, seeking to entice us into craving them for a lifetime. That's their job. And regardless of the source, our internal feedback agents—the taste and olfactory receptors throughout our bodies—keep detecting and reporting tastes and flavors. But they can't always distinguish a flavor composed of a well-made fake from flavors rooted in phytochemicals and essential fatty acids and amino acids.

Certain flavors are go-to's in engineered food products. Recall what Fred Provenza's tannin experiment with goats revealed. One day the goats found the tannin-containing pellets appealing. The next day they shunned them. This is not the behavior flavorists seek to cultivate. Bitter tastes are neither reliably seductive nor addictive. It's the gateway tastes—sweet, salty, and umami—that get people to try new products and hanker for more.

With sugar being relatively rare in our ancestral environments, we nabbed sweet foods whenever we could. Sweet tastes work particularly well to motivate children to eat foods. Just visit the cereal aisle at the grocery store. Have you ever overheard kids haranguing their parents to buy the kind advertising whole grains? Nope. They want the sweetened ones. Kids have no idea that the flavor-feedback relationship is at work, but they do know they like eating dessert for breakfast.

Oleogustus has gateway qualities too. There is hardly a person on the planet who turns their nose up at the way a well-deployed fat creates a perfect French fry. The same goes for the old standby, salt. It is the rare soul who can eat just one crunchy, salty chip when gathered with friends and family to watch the Super Bowl.

The way body wisdom responds to flavor not only molded the human diet of the past; it continues to do so today. Consider the remarkable success in recent years of faux meat, cheese, and dairy products. These ultra-processed foods are heavy on palatants and other appetite-stimulating additives. The better ones grill, melt, spread, and pour much like the real thing. Subliminal imagery on the packaging evokes farm scenes, some quaintly reminiscent of an heirloom seed packet or a pastoral landscape. Compelling eco-friendly and health-focused product narratives help sales for targeted demographics. So does putting faux animal foods shoulder to shoulder with real ones on grocery shelves and calling them by the same name—meat, cheese, milk, and butter.

But what, really, do ersatz meat and dairy deliver? There are key differences in the nutrient profile of ultra-processed food products and foods made from common, centuries-old methods of food processing like milling grains to make bread, cooking berries to make jams, or turning tomatoes into sauce. Ultra-processed foods are fundamentally different in several ways. First, they rely on ingredients usually obtained from low-cost, industrial sources that are extracted from whole foods—like the protein isolates from peas and soybeans used in faux meats. Generating such ingredients requires slicing, dicing, heating, and otherwise deconstructing whole plant foods. This level of processing, beyond the capacity of a home kitchen, changes the composition and amount of health-imbuing flavor compounds, notably phytochemicals and fats.

Another hallmark of ultra-processed foods, as you might imagine, is that a pile of protein isolate tastes wretched and looks just as bad. This is where palatants and other substances come in: they take the ingredients that go into ultra-processed foods from barely edible to hyperappealing using the most favored tastes—sweet, salty, and umami. Industrial-scale machinery easily turns a pile of mush into a more pleasing and familiar shape through extruding a food product into final form, whether patty, sausage, or loaf. Lastly, texture too plays a role in flavor. Most of us crave crunchiness, another feature of many ultra-processed foods. While we all need to eat, no one really needs to consume ultra-processed food.

It is challenging if not impossible to re-create the full nutritional profile of the whole foods from which ingredients for ultra-processed foods are sourced. But this is not the goal. Adding back the same phytochemicals and fats removed in processing would wreck the formulated textures, tastes, and flavors of ultra-processed foods. Much easier is to begin with a blank-slate pile of protein or carbs and add vitamins and minerals back into the final version of a food product, along with various flavorings, binders, additives, and preservatives. But whatever the final concoction, it is not a substitute for all that was lost along the way.

Tastier foods are no longer a sure guide to healthier ones. Flooding ultra-processed foods with flavors may delight the tongue and brain for a moment, but it leaves body wisdom to spin aimlessly like the needle of a broken compass searching for north. Lost and on our own, we fish for nourishment in a treacherous sea of engineered flavors. It seems we have unwittingly turned ourselves into a two-legged version of cows eating TMRs from a fancier trough.

Getting in Touch

The next time you have a tasty whole food of some sort nearby, pinch your nose shut and take a bite. How did it taste? Probably there wasn't much if any taste at all. When you closed your nose, the flow of volatile-laden air had nowhere to go, so it bottomed out in the dead end of your mouth, leaving your olfactory receptors high and dry. For that brief moment your nose was lost, little more than a useless nubbin of flesh sticking out from the middle of your face.

Now take another bite, but don't pinch your nose. With the air flowing, the compounds that got volatilized in your mouth can once again reach receptors in the hinterlands of your nose. In far less time than it takes you to read this sentence, your olfactory receptors forward information about the volatiles upstairs to the brain. As the final arbiter in such matters, the brain registers them as a flavor and files it away for future reference. Perhaps the source of the flavor-conveying volatiles was a really peachy peach, an extra-bitter Brussels sprout, or an inviting, umami-packed tomato soup. This is the type of intelligence you want filling the nutritional dossier on which body wisdom feeds, for it codes for compounds and nutrients that underpin health.

The same hack works with beverages too; just make sure to pick one laden with phytochemicals, like coffee, tea, or wine. Take in a small sip, give it a few gentle chews to release the volatiles, and let it sit in your mouth a moment before slowly swallowing. Chewing acts like a bellows and floods your olfactory receptors with the volatiles that your brain can then register as more complex and satisfying than if you swallowed quickly. That's what all the sloshing and jaw movements are about as a sommelière gets into tasting her wine.

Over the long span of human history the intertwining of agriculture and body wisdom led cultures around the world to develop regional and local cuisines. Although the dishes of each cuisine are unique and distinct, they share deep roots. The reason is simple. Every cuisine is steeped in combinations of ingredients sourced from an ancient palette of flavor-making compounds and nutrients found in whole plant and animal foods. Historically, those that kept people healthy were the ones that stuck. So the next time you sit down to enjoy a meal made from foods with genuine flavors, think about the tastes and flavors like signposts that help your body wisdom steer clear of potholes and collisions.

That the health effects of diet are poorly understood through single components is borne out in research showing that combinations of various foods and spices can prove more important than the effects of any individual food or spice. For example, data from the Iowa Women's Health Study show that fiber left intact in whole-grain foods reduced the risk of mortality more than an equivalent amount of fiber added back into to refined-grain foods. The benefit wasn't just from the fiber. Eating whole foods sets up synergies between basic components like fats, fiber, and phytochemicals when they come naturally packaged together in a plant.

Another example of dietary synergies comes from the well-known Mediterranean diet, which is associated with lower risk of heart disease, diabetes, and breast, colon, and prostate cancer. Tomatoes, common in many Mediterranean dishes, contain lycopene, a fat-soluble carotenoid that cooking makes more bioavailable. The combination of cooked tomatoes and a fat (olive oil) increases lycopene's antioxidant effects, which inhibit inflammation and promote the death of tumor cells. Naturally, preventing oxidative damage in the first place beats trying to repair the damage it causes. Seen in this light, phytochemicals shine as preventive medicine.

Carotenoids and vitamin E in meat protect proteins and highly unsaturated long-chain omega-3 fats from oxidation. This effect, along with the contribution of phytochemicals to distinct flavor profiles, is part of the dual benefit of livestock eating a diet rich in different living forages. So while you might expect the higher omega-3 content of grass-fed beef to confer shorter shelf life, it actually stores better. Terpenes, phenols, carotenoids, and other antioxidants also occur at higher levels in the meat and milk of livestock that graze diverse pasture forage, conveying flavor on which our body wisdom can act. And if garlic or oil from juniper, rosemary, or cloves gets added to the diet of sheep or cows, it influences the phytochemical profile of meat, reducing oxidation and enhancing flavor.

Drinking polyphenol-rich red wine along with meat is another way to get some phytochemicals in your diet. This food and drink combination was also found to greatly reduce blood levels of key markers for oxidative stress and inflammation related to heart disease. These effects are likely due to polyphenols inhibiting the oxidation of dietary fat and beta-carotene. In simplifying our modern diet we've stripped out things that formerly interacted to our benefit. It's in our interest to rethink the dietary relevance of phytochemicals, much like we did for fats in moving beyond the legacy of Ancel Keys.

Despite the collective scientific and practical efforts that have uncovered how conventional farming practices compromise the mix and abundance of phytochemicals, fats, and other compounds in food, we still have an incomplete understanding of their synergies and effects on health. Nonetheless, we know they benefit us through an ensemble effect, proving more potent and versatile as a team. Combinations of particular phytochemicals can even enhance the effectiveness of conventional antibiotics against laboratory cultures of multidrug-resistant bacteria and food-borne pathogens. Therein lies yet another example of the inherent nutritional wisdom of whole food–based cuisines.

Still, it remains challenging to conduct dietary research. One factor is that most studies based on food frequency questionnaires do not actually measure food intake or the nutritional composition of what study subjects actually ate. Instead nutritional profiles are assigned to specific foods from values in published compositional tables that may not accurately reflect what a person ate, let alone what their food ate.

An extreme example of how food frequency questionnaires can skew results comes from a study that found self-reported caloric intakes were inadequate for human survival for more than a third of the participants in the National Health and Nutrition Examination Survey. Such underreporting is a key problem of many diet-related epidemiological studies and has led some critics to go so far as to contend that research using self-reported information stands on little more than guesswork.

However, a 2019 study in which food and diet were actually measured and quantified reveals some interesting insights. Investigators at the National Institutes of Health published results from a controlled, randomized clinical trial that compared the effects of an ultra-processed diet to a diet of unprocessed foods.

Twenty initially stable-weight adults (ten men and ten women) lived at a clinical center for a month and ate a diet of either ultra-processed foods or unprocessed foods for two weeks and then switched to the other diet for two weeks. Subjects chose what to eat from a menu of calorically and nutritionally matched meals and snacks. In terms of nutritional composition, the diet of ultra-processed foods had far less fiber and more total sugar, saturated fat, and omega-6s, whereas the diet of unprocessed foods had little sugar, a lower ratio of omega-6 to omega-3 fats, and far more fiber. Participants were free to eat as much or as little as they wanted on each diet. Providing all the subjects' food allowed the researchers to reliably track everything they consumed and avoided the problem of people inaccurately reporting what they ate.

The study subjects ate about 500 more calories a day on the diet of ultra-processed foods compared to when they ate the diet of unprocessed foods. Unsurprisingly, the additional caloric intake led to weight gain, an average of a pound a week. In contrast, when on the diet of unprocessed foods, participants lost a pound a week. In other words, when on the diet of ultra-processed foods, participants ate well beyond when they were full.

A comparison of biomarkers in the blood of study subjects at the end of each diet period revealed further insights. When on the diet of unprocessed foods, they had significantly lower cholesterol levels and higher levels of hormones associated with satiety and glucose regulation. It seems that a diet rich in ultra-processed foods is an effective setup for changing one's metabolism—but in the wrong direction.

Key lessons have emerged from decades of research in nutrition, notably that dietary habits are linked to a wide range of chronic conditions, that focusing on dietary patterns works better than keying in on single nutrients, and that prevention is more effective than treating chronic diseases and obesity. Breeding the taste out of foods and manipulating our body wisdom with ultra-processed foods flooded with palatants did more than make healthy foods bland and unhealthy foods tasty. It displaced foods rich in micronutrients, phytochemicals, and healthy fats from the American dinner table.

Dietary Dilemmas

In the early to mid-twentieth century nutritional science identified deficiencies in single nutrients as the cause of diseases like pellagra and rickets. That low levels of a single nutrient could manifest as an identifiable disease was a powerful paradigm. Such deficiencies were relatively easy to correct and made dietary advice simple and convenient. By the century's close, diseases resulting from once-common, single-cause dietary shortfalls virtually vanished from American lives.

This remarkable improvement in public health set the stage for investigating the growing epidemic of chronic diseases through the same lens. It also launched dietary policy on a trajectory focused on supplementing foods with individual vitamins and minerals. But as dietary recommendations zeroed in on single-nutrient solutions, like reducing consumption of fats, obesity and associated chronic diseases spiraled out of control. Focusing on singular causes and seeing broad classes of nutrients as winners or losers obscured fundamental differences between different kinds of carbohydrates and fats and the importance of phytochemicals. Terms like *low-carb* or *high-carb* fail to distinguish between a simple refined carbohydrate, like the sugar in the frosting on a cake, and the more complex ones in whole fruits, vegetables, and grains that come naturally packaged with phytochemicals, vitamins, and minerals.

The rush to low-fat diets fueled explosive growth in reduced-fat products rich in refined carbohydrates of dubious nutritional value. As Americans followed dietary advice to shun fats, they chowed down on low-fiber carbs and their bodies blew up like balloons. Today three-quarters of Americans consume an abundance of refined grain products promoted as low-fat foods.

One area of woefully oversimplified dietary guidance is the standard advice to choose lean over fatty meat. Low-fat meats are not intrinsically better healthwise. Processed meats formulated to be low fat also tend to contain lots of salt and additives like nitrates, things best eaten in minimal amounts. And as we've covered, much depends on the specific fats we eat. Grass-fed animals generally deliver more omega-3s, CLA, vitamins A and E, and phytochemicals with antioxidant and anti-inflammatory properties. We could all benefit from more of these things in our diets. While some epidemiological studies associate high red meat consumption with an increased risk of heart disease, cancer, and diabetes, such studies have yet to take the diet of the animals into account, let alone what effect this has on our health. Dietary policy generally sidesteps this problematic wrinkle.

As we're continuing to learn about the health ramifications of crops with low levels of beneficial phytochemicals and other bioactive compounds, bland fruits and vegetables advertise how conventional agriculture broke the bond between flavor and health. When we deprive our taste and flavor receptors of edible intelligence, we rob our bodies of vital information central to navigating nutritional seas.

Another key shortcoming in national dietary policy is that beyond "eat the rainbow" style advice, the benefits of a phytochemical-rich diet remain underappreciated. Guidance to eat more whole plant foods somehow morphed into many people thinking that food and beverages made from ultra-processed plants are nutritionally equivalent. This is troubling given the evidence that our bodies benefit most when phytochemicals and other bioactive compounds land on our plates in the mix and proportions in which they occur in whole foods.

But before you start purging your cupboard of all those omega-6-rich oils extracted from corn, sunflower, and other seeds, remember that the role that omega-6 fats play in firing up inflammation is normal. Characterizing omega-6s as "bad" falls into a pattern of thinking that assumes these fats have only one function in our bodies and in the plants or animals from which we get our foods. The way that fats in the human diet interact with our bodies is more nuanced, complex, and context dependent—far more.

We need to think about both the total amounts *and* the balance of LA and ALA. Remember, they are both *essential* fatty acids and precursors of longer-chain fats in the human diet. And unlike ruminants our bodies cannot make fat directly from the land. For better and worse, the range and diversity of fats in our bodies comes from what our food ate.

BALANCING ACT

Tell me what you eat, and I will tell you what you are.

—JEAN ANTHELME BRILLAT-SAVARIN

Of the many people since Hippocrates to recognize connections between diet and health, one of the most celebrated is the French lawyer who wrote a famous (though not particularly readable) book about food. Jean Anthelme Brillat-Savarin was born in 1755 to a bourgeois family in the town of Belley in eastern France and had the great misfortune to become mayor of his hometown a year before the French Revolution. In 1793, he fled on foot across the Alps and then on to the United States. A few years later he returned home to serve the Republic as an appeals court judge, writing extensively on politics and the law. But he is remembered for popularizing gastronomic writing in his 1825 memoir that gave us a famous aphorism shortened in translation to "you are what you eat." Today, his name stands immortalized in a delicious French triple cream cheese that's more than three-quarters fat.

In a series of 28 meditations he focused on discourses around the pleasures and aesthetics of the table, the beauty of a simple meal well prepared, and thoughts on cooking. He also asserted insights into the role of diet in health. In talking with obese people Brillat found many expressed strong cravings for sugary or starchy foods—bread, pasta, rice, potatoes, and sweets. He thought obesity an unnatural state, noting how neither carnivores nor grazing animals grew overly fat in the wild. But when fed on grains or potatoes, predator and prey alike bulked right up. Seeing a pattern in all this, he held that a diet rich in sugar, starchy foods, and white flour paved the way to obesity.

To maintain health, he recommended a diet rich in meats, leafy greens, root vegetables, and fruits—and light on starchy carbohydrates like bread and pasta. In other words, a diet resembling the ancestral human diet. He asserted that those who followed his dietary advice would enjoy vigor and health, feeling fresh and energetic through life's grand adventure.

Over subsequent centuries, advances in nutritional science shaped and reshaped conventional dietary wisdom and recommendations. Single nutrient deficiencies came into focus after Casimir Funk proposed in 1913 that unprocessed rice husks contained a "vital amine" that could protect chickens from beriberi. By midcentury, efforts to identify, isolate, and synthesize new vitamins that could prevent disease delivered antidotes to common maladies. Vitamin B_1 worked like magic to clear up beriberi. So did other vitamins—A for night blindness, B_3 for pellagra, C for scurvy, and D for rickets. Public health efforts increasingly focused on fortifying staples like milk, cereals, and bread with vitamins and minerals.

Throughout human history the specter of starvation defined the primary enemy in our relationship with food. Quantity remained the agronomic goal. Over the past century, an emphasis on ensuring adequate calories encouraged producing cheap, energy-rich, highly processed foods that store and ship well. The American diet followed those of chickens and cows down the road of progress, replacing fresh, whole foods with calories from refined grains containing fewer micronutrients, phytochemicals, and beneficial fats. Today, however, the main problem across much of the world reflects inequitable access to nutritional quality as much as inadequate calories.

Diet-related chronic diseases were surging in wealthy countries, and obesity rates were climbing by the closing decade of the twentieth century. In 1980, the Centers for Disease Control estimated that just under half of Americans were obese or overweight. By 2010 that figure rose to nearly three-quarters. Subsequent dietary recommendations remained largely focused around single nutrients and the presumed benefits of a low-fat diet. Avoid cholesterol. Don't eat saturated fats. Such authoritatively cast advice fueled demonization of meat and dairy products. It also spawned an ever-growing selection of highly processed food products rich in refined carbohydrates to meet growing demand for low-fat foods. Our bodies made their own plans for dealing with all the simple sugars: turn them into fat and store them around our middles or tuck them into organs like our liver. It seems low-fat diets proved rather fattening and unhealthy for a lot of us.

In addition, newer science points to the importance of an individual's microbiome in digestion, metabolism, and other aspects of health. Weight regulation, for example, is no longer considered to simply reflect how much one eats and exercises. The human microbiome also influences how much energy (calories) we get from the foods in our diet, as well as a vast array of compounds that affect other aspects of human biology.

An increase in obesity and the constellation of ailments that stem from it was not the outcome nutrition professionals sought in backing a dietary shift away from fats. As linkages between cholesterol and health became clearer, they also grew more complicated. Two types of cholesterol—low-density cholesterol (LDL cholesterol) and high-density cholesterol (HDL cholesterol)—differ in their effects. The so-called "bad" type of cholesterol (LDL) can stick to and accumulate on the inside of arteries, narrowing their size enough to reduce or obstruct normal blood flow, increasing the risk of heart attack or stroke. In contrast, the "good" kind (HDL) also travels around in blood, but instead of lodging where it shouldn't, it picks up and delivers the bad type to the liver for disposal.

The ways the two main kinds of cholesterol interact with the complexity of human biology helps explain the shortcomings of using total cholesterol as the main predictor of cardiovascular disease. Through the 1990s more studies and evidence chipped away at the overly simplistic fat-is-bad-for-you misconception.

Two such studies occurred in our hometown of Seattle, where researchers from the University of Washington recruited hundreds of men and women from the ranks of Boeing. Study subjects all had high levels of LDL cholesterol, and some had high levels of triglycerides, another fat biomarker used to assess the risk of cardiovascular disease. Both men and women were put on a range of low-fat diets to see which one might prove most effective at lowering LDL cholesterol and triglyceride levels. A year on the different low-fat diets produced mixed results. The subjects' LDL-cholesterol levels went down. But so did their HDL cholesterol, the good type. And triglycerides had gone up, another undesirable outcome.

A follow-up study that explored gender differences found that in women the decline of good cholesterol, the HDL fraction, was so great that it increased the risk of cardiovascular disease. Several additional studies confirmed the awkward truth that low-fat diets lower the good kind of cholesterol, especially in women.

This gender difference is of particular concern given that women diet more than men. And with so much dietary guidance framed around cutting back on fats to lose weight, that's what a lot of women did. Yet we now know that adequate levels of certain fats, like omega-3s, are vitally important.

The basis for supporting a low-fat (and thus high-carb) diet crumbled as research marched on. A variety of studies that followed health outcomes or fed people supervised diets found that some fats produced health benefits and that overconsumption of sugary and starch-rich foods consistently proved detrimental to health. About a decade after the studies on Boeing employees were published, an analysis of cohort studies evaluated the association of saturated fat with cardiovascular disease. The authors succinctly summed up their findings in reporting they found no significant evidence that dietary saturated fat increased the risk of heart disease or fatal heart attacks.

This was pretty much the same conclusion the Food and Nutrition Board of the National Academy of Sciences arrived at some 30 years before. After reviewing the available evidence for and science behind the benefits of a low-fat diet, the board found that there was scant evidence for urging Americans to kick fats out of their diet. Their *Toward Healthful Diets* report stirred up substantial controversy in Congress because it pretty much recommended the opposite of the newly released, first-ever U.S. Dietary Guidelines. The media had a field day with the mixed messages.

A Good Fat

One case for which the health effects seem clear is CLA, the fat ruminal bacteria make that becomes part of milk and meat. Changing the diet of cows to grains, concentrates, and the like coupled with the rising popularity of low-fat dairy products, dramatically decreased consumption of CLA across the Western world. How much should we care? Few molecules in the human diet deliver the wide range of benefits that CLA does. In animal, laboratory, and human studies CLA exhibits potent anti-inflammatory activity and has been found to help prevent certain types of cancer, type 2 diabetes, and atherosclerosis (plaque buildup in arteries). Conjugated linoleic acid also appears to help modulate immune system functions, suppressing inflammation associated with, and even showing promise for potentially treating, inflammatory bowel disease and colorectal cancer. And, counterintuitively, it has been shown to reduce body weight in obese individuals.

Yet our bodies have limited ability to convert dietary precursors into CLA. We churn out nowhere near the amount a healthy cow does in her milk. For a fat as beneficial as CLA, every bit counts—and dairy products and meat are the only natural source of it.

It wasn't until the late 1970s that scientists really began to investigate CLA. One in particular, Michael Pariza at the University of Wisconsin, stumbled onto something unexpected while cooking up hamburger to study the effects of cooking time and temperature on the formation of mutagenic compounds. While looking for bad compounds, he discovered a good one—something that inhibited the formation of cancer-causing substances. Intrigued, he carried out a series of experiments on mice in which some received topical doses of the mystery substance immediately before skin tumors were induced. The dosed mice developed half as many tumors as the control group. By 1987 Pariza and his colleagues identified the tumor-inhibiting substance. It was CLA. Another decade passed, and in 1996 the National Research Council found that evidence from animal research unequivocally demonstrated the anticancer effects of CLA.

Since this early work much more has been learned about CLA. Nearly 30 different forms (isomers) have been discovered. The two chief ones—rumenic acid and *trans*-10, *cis*-12—turn out to have significant and different health benefits.* Rumenic acid is the most abundant form of CLA, and levels are generally higher in milk than meat.

Animal models and human studies have continued to show that CLA influences a wide range of anti-inflammatory gene pathways and

* CLA nomenclature stems from the location and orientation of double bonds and distinguishes various isomers from one another. Rumenic acid is the *trans*-9, *cis*-11 isomer. The numbers after "cis" or "trans" relate to the location of double bonds, and *cis* and *trans*, respectively, indicate whether the double-bonded hydrogen atoms are on the same or opposite sides of the carbon chain. A *cis* bond creates a bend in the chain, whereas a *trans* bond produces a straight section.

enzymatic activity at the cellular level. In one case, consuming butter rich in rumenic acid increased serum levels of an anti-inflammatory molecule (interleukin 10) that helps modulate inflammation in overweight individuals. The other common form of CLA, the *trans*-10, *cis*-12 isomer, is thought to help our bodies combat cancer, obesity, and diabetes, as well as inhibit plaque formation on the walls of arteries.

Studies on using various isomers of CLA to treat breast and colon cancer have produced mixed results, but some have been shown to help quell inflammation, which would contribute to preventing cancer in the first place. Moreover, CLA can cut off angiogenesis, the process tumors deploy to pirate their own dedicated nutrient supply from nearby blood vessels. CLA also helps thwart cancer through promoting apoptosis, the body's programmed cell death process for eliminating abnormal cells (like cancer).

Interestingly, CLA in meat is concentrated in the interstitial, nonvisible fat distributed along muscle fibers. This is not the marbled stuff that grows in feedlot cows. It is the type of fat animals put on when they walk around grazing.

A 2001 study of more than 5,000 men evaluated mortality over 25 years and found no evidence that daily milk drinkers faced a greater risk of death from any cause, let alone heart disease. To the contrary, greater milk consumption was associated with a lower risk of dying from heart disease or cancer. Similarly, a U.S. study of more than 3,000 young adults found that milk consumption appeared to not only protect overweight individuals from developing obesity but also reduced the odds of developing insulin resistance and cardiovascular disease. A study of several thousand two-year-old Dutch children found the incidence of asthma to be lower among those who consumed whole milk or butter on a daily basis. Finally, a 2019 overview of 42 prior reviews of the health effects of dairy consumption found mixed evidence for the risk of various types of cancer, with strong evidence that it reduced the risk of colorectal cancer.

Missing from all this research, however, is that all milk is not equally healthful. Although the diet of a cow is the single biggest factor in setting the type and levels of saturated fats, omega-3s, and CLA in milk and meat, dietary studies generally don't account for differences in what the cows ate.

Ancestral Fats

The recent rise in chronic diseases and metabolic disorders does not reflect changes in our genes but in how our genes respond to the Western diet and its abundant calories and scant amount of fiber, micronutrients, and phytochemicals. And of course the kinds and amounts of fat also figure into our body's response to the modern human diet.

As it turns out, the earliest evidence of meat and fish in our ancestral diet falls around two million years ago, coinciding with the onset of gradual increases in brain size. The East African Rift Valley, where our pre-human ancestors lived and evolved, is dotted with large freshwater lakes, and early evidence of sophisticated tool use relates to aquatic foods. That the ratio of omega-6 to omega-3 of many tropical freshwater fish and shellfish is around 1:1, close to what our bodies maintain in the human brain, led scientists to suggest that increased consumption of omega-3s helped drive changes in brain size. But perhaps aquatic foods were not the sole factor at play as wild game, which our ancestors also ate, has close to a balance of omega-6s and omega-3s.

Many millennia later, with the advent of agriculture in the post-glacial world, our diet changed again as grains became a dominant source of calories for many, and milk became available across Europe, India, and Africa through domestication of livestock. In these regions the ability to digest milk and dairy became more common in human populations that took up dairy farming early in the agricultural age.

More recently, we've embraced another major dietary shift related to fat consumption. As we've covered, over the past century people in Western countries began eating far more omega-6-rich processed foods, and changes in agricultural practices decreased the omega-3 content of our primary dietary sources—meat, milk, and eggs. The abundance of omega-6 fats in the Western diet is now estimated at between 10 and 20 times higher than omega-3 levels.

In the half century from 1935 to 1985, total fat intake in the American diet increased by almost a third, with vegetable oils accounting for most of the increase. Since the 1950s, consumption of omega-6 fats in Western diets almost doubled while omega-3 intake cratered. Omega-6-rich oils now account for almost a tenth of the calories in a typical American diet. From 1980 to 2005, the average ratio of omega-6s to omega-3s in human breast milk roughly doubled in the United States (to about 16:1) and the ratio of omega-6 to omega-3 fats in our bodies rose to over 20:1. We've turned supermarkets into feedlots for people.

This dramatic rise in the ratio of dietary omega-6s to omega-3s is thought to have exacerbated or predisposed people to inflammatory diseases through shifting immune responses toward states of chronic inflammation. We've inadvertently run an uncontrolled experiment on ourselves, radically shifting the chemical makeup of the fats that have long underpinned human biology and health, especially our immune system.

Resolutionary Act

Inflammation is something we definitely want our bodies to be able to do. It gives our immune system spectacular defensive capability—killing pathogens, healing wounds, and halting abnormal cell growth. But it is also double-edged as healthy bystander cells get caught in the crossfire of an inflammatory event.

Normally inflammation is a short-term process, so over a lifetime cells and tissues recover. But when it becomes a chronic condition, the repeated dousings of caustic compounds that immune cells release during inflammation can seed maladies. Unending inflammation in the brain is considered a big factor in Alzheimer's, while heart disease has its roots in inflamed arteries. Asthma arises from inflammation in the respiratory system. Ailments of the digestive system like Crohn's disease and irritable bowel syndrome also have been linked to inflammation. It seems no part of our body is completely immune from the immune system itself.

Since the time of Ancel Keys, the understanding of how fats influence our immune system has grown considerably. Certain fats, it turns out, start inflammation while others bring on the end of inflammation—in particular key ones we inadvertently stripped from our diet. Arachidonic acid, the omega-6 fat, is on the front end of inflammation. A few tweaks to its chemical structure and it morphs into eicosanoids. These inflammation-starting molecules also regulate other immune cells, notably those that create "cytokine storms." This condition, akin to an immune system stuck on overdrive, led to death or severe symptoms among many COVID-19 patients.

Recall that cell membranes are where our bodies stockpile both omega-3 and omega-6 fats. So if one's diet is abundant in foods rich in arachidonic acid (or other omega-6 fats readily converted to it), your immune system will always have what it needs to fire up and sustain inflammation. This is perfect for fending off a virus, fighting incipient cancer, or nursing a wound every once in a while. Once these problems are gone, inflammation needs to stop—pronto. But when it doesn't, the situation becomes like a stove left on day and night until the house burns down.

Until quite recently immunologists thought inflammation ended passively with omega-6-derived eicosanoids and similar molecules simply dissipating over time. This ran-out-of-steam hypothesis is a nice, tidy idea. But it doesn't square with another reality—the prevalence of diseases among Americans linked to chronic low-level inflammation. A substantial body of evidence implicates a high ratio of omega-6 to omega-3 as contributing to the inflammation that underlies obesity and chronic diseases like cardiovascular and irritable bowel disease, rheumatoid arthritis, and neurodegenerative diseases such as Alzheimer's.

The reality is that our immune system dials back on inflammation through a process as complex and orchestrated as its initiation. The trio of long-chain omega-3s (EPA, DHA, and DPA) are precursors for the barn door–closing molecules that end inflammation, but only if there are enough of them stockpiled in cell membranes. In a refreshing bit of biochemical nomenclature, the names of these inflammation enders—resolvins and protectins—say what they do. They are among the molecules that spur immune cells on the front end of inflammation to wind down and reverse course.

However, there is a curve-ball aspect to ending inflammation. Eicosanoids built from omega-6s play a role in ending inflammation as well as starting it. How so? Bring in the right enzyme, and eicosanoids rapidly morph into lipoxins, a cousin-like group to protectins and resolvins. Once all these fat-related molecules are in place, voilà! The beginning of inflammation transforms into its end, returning the body back to where it wants to be—at peace with itself, humming along in homeostasis.

There is a larger point embedded in the role of omega-3s and omega-6s in inflammation too. With the exception of organ transplants and the like, where immune system suppression is necessary, routinely dampening the immune system with anti-inflammatories can prove as bad as chronic inflammation. Hamstringing a normal immune response reduces the efficacy of vaccines, as well as our ability to fend off pathogens, heal wounds, and eliminate cancerous cells. For your body to pull off the balancing act that inflammation really is—not too much and not too little—you need to have a well-stocked balance of fats to draw on.

Balanced Stores

Although our bodies can convert some fats within the same family (as among long-chain omega-3s), we can't convert omega-6s into omega-3s. We lack the gene and thus the enzyme needed for an otherwise straightforward bit of biochemistry. This simple reality means that what we eat controls the amount and balance of these two types of fats in our bodies. The amounts circulating in our blood change over the course of days based on what our food provides. Once incorporated into our bodies, fats can stick around for months in our cell membranes, stored until needed for other uses. So changing the types of fats in one's diet shapes the stockpile of omega-3s and omega-6s our immune system draws on.

So what foods are full of the longer chain omega-3s? We've already covered that wild cold-water fish like salmon are good sources. Others include mackerel and smaller fish like anchovies and sardines. Plants, with the exception of some seaweeds and algae, lack the longest-chain omega-3s. However, walnuts, broccoli, and the chloroplasts of green leafy vegetables are all rich in the essential omega-3 (ALA) as are a few seed oils, flaxseed and canola among them. Our bodies can use these ALA sources to make the longer ones, provided omega 6s aren't hogging the shared biochemical pathway. There are also other types of beneficial fats. Olive oil and avocados have a lot of oleic acid, a mono-unsaturated omega-9 fat with anti-inflammatory properties, and are rich in antioxidant phytochemicals as well.*

And as we've seen, the abundance and balance of omega-3s and omega-6s in meat, dairy, and eggs are directly linked to their levels in the foods that compose an animal's diet. This matters because along with seafood, livestock account for the vast majority of long-

* Monounsaturated fats, you may have deduced by now, have only one double bond along their carbon chain.

chain omega-3s in the American diet. So when animals get shorted on omega-3s, their dearth becomes ours, rippling through our diet and into our bodies.

A range of studies have looked at health effects of omega-3s and omega-6s singly and as a ratio. Greater consumption of omega-3 fatty acids has been shown to decrease cholesterol levels, mortality from heart disease, and the effects of various inflammatory and autoimmune diseases. Specifically, increased dietary omega-3 consumption (or supplementation) also has been shown or thought to reduce the effects or risk of Crohn's disease, arthritis, asthma, diabetes, and obesity. And animal studies show that increased omega-3 consumption can reduce weight gain and body fat, slow prostate cancer growth, and dial down the inflammation that underlies colon cancer.

While clinical evidence of treatment efficacy remains highly variable between studies, greater dietary consumption of omega-3s has been associated with lower inflammatory markers in people. For example, a study of over a thousand Italians spanning the full range of adulthood (ages 20 to 98) found those with the highest omega-3 levels had the lowest levels of inflammatory markers, while those with the highest omega-6 levels had the highest levels of such markers. Several reviews of randomized controlled trials of omega-6 intake (LA) likewise found more inflammatory markers at the highest dietary intake. However, the same researchers reported little evidence for dietary ALA (the essential omega-3) intake affecting inflammatory markers in a review of randomized controlled trials.

Reviewing these studies it appears that a wide range of baseline omega-3 and omega-6 consumption and variable health conditions and levels of obesity among participants is partly responsible for disparate findings. A 2008 review of randomized controlled trials concluded that while the absolute amount of long-chain omega-3s was relevant to dietary effects on the risk of cardiovascular disease, the ratio of omega-

6s to omega-3s was not. Two years later, however, a study of blood levels of omega-3s in several hundred healthy men and women found that higher concentrations were associated with lower inflammatory markers and that the strongest correlation for inflammatory markers was the ratio of omega-6 to omega-3 fats.

The takeaway from these studies is that it doesn't always enhance understanding to look at fats in isolation. We not only need adequate amounts of both omega-6s and omega-3s in our diet, but we also need these fats in relatively balanced amounts. Although the ratio does not appear to matter much when intake rates are low and both are in short supply, at high consumption the ratio appears to really matter.

How much did the effect of a diet with a lopsided balance of omega-6 and omega-3 fats amplify the toll of COVID-19 in Americans? A 2020 study cited evidence that long-chain omega-3s (EPA and DHA) could control cytokine storms. Whether or not dietary treatments ultimately prove useful, medical researchers have suggested that inadequate levels of omega-3s, micronutrients (like zinc), and antioxidants all played roles in poor health outcomes for individuals who contracted COVID-19. And as we've discussed, the way we grow crops and the diets we provide livestock profoundly influence how much and what types of fats, minerals, and phytochemicals end up in our bodies.

Atomic-scale imaging of the part of the spike protein that the COVID-19 virus uses to enter human cells revealed linoleic acid (LA, the essential omega-6) in its structure. Medical researchers hypothesized that access to ample amounts of LA in infected people may enhance the ability of the virus to enter cells and replicate. Though speculative, this scenario is possible. After all, it's well known that the fat composition of a cell membrane can affect the ability of a virus to fuse with it and thereby influence viral replication. For example, an omega-3-derived protectin inhibits influenza virus replication, thereby improving patient resilience to severe bouts of the flu. Moreover, the resolvins and protectins built from long-chain omega-3s are potent regulators of severe inflammation in airways. If similar connections also influence COVID-19, then what one ate over the past few months could affect one's susceptibility to symptomatic or severe infection.

Other evidence suggests that a diet with balanced omega-6s and omega-3s is best for overall human health too. Omega-3s have been studied in the context of specific types of cancer and have been found to both reduce breast cancer risk and inhibit growth of breast cancer cells. A study of biopsies from several hundred French women found that those with the highest omega-3 levels had less than half the risk of breast cancer. Clinical studies likewise find higher omega-3 consumption associated with lower breast cancer risk.

Long-chain omega-3s also have been shown to suppress growth of prostate cancer cells in cultures by interrupting chemical signaling that stimulates tumor growth. This may help explain why a study of Icelandic men found that those who consumed fish oil late in life had roughly half the risk of advanced prostate cancer. In addition, several studies of prostate cancer patients found those consuming the most fish died at about half the rate of those who ate the least. Other studies report comparable reductions in the risk of colorectal cancer among older men who regularly consumed fish or fish oil supplements.

Higher consumption of long-chain omega-3s also has been associated with a reduced risk of cardiac arrest and sudden death. Meta-analyses of controlled intervention trials found that fish and omega-3 consumption reduced inflammatory biomarkers for heart tissues, improving heart function and lowering blood pressure. One of the most recent, a 2021 analysis of 17 cohort studies that followed tens of thousands of individuals for an average of 16 years, found that those with the highest levels of one or more of the trio of long-chain omega-3s in their blood had a significantly lower risk of premature death. And while studies report conflicting results on fish oil supplementation, there appear to be no adverse effects: although it's safe for everyone, it may only benefit some.

While it seems that both men and women could reduce their risk of particular types of cancer through eating more omega-3-rich foods, some well-publicized studies found little benefit of supplements. For example, reviews of randomized, double-blind, placebo-controlled trials on the effects of omega-3 supplements on cardiovascular disease found no evidence of a preventive effect. Likewise a review of randomized controlled trials found little evidence that omega-3 supplementation influenced overall cancer incidence or mortality. It seems the jury is still out as to the value of supplements compared to increasing omega-3 intake in whole foods.

Yet studies that tightly controlled for long-chain omega-3 intake have documented relationships between blood levels of EPA and DHA and the amount consumed, whether in food or in dietary supplements. Eating more of these fats raises their levels in blood, although it takes about a month for blood plasma concentrations to equilibrate to regular supplementation. In addition, dietary intake of EPA and DHA reduces blood plasma concentrations of the long-chain omega-6 arachidonic acid (ARA). And a seven-day experiment with dietary supplementation of DPA and EPA found increased blood plasma concentrations of various inflammation-quelling resolvins.

In moms-to-be, the EPA and DHA content of their breast milk is directly related to what they eat. Of course, not all infants are breast-fed, and those who aren't generally receive formula made from plant oils. But maybe they shouldn't. In a double-blind, randomized trial healthy newborns received formula including either dairy fats or plant oils. After four months the red blood cells of the group receiving dairy fats had 50 percent more omega-3s, a quarter more DHA, and almost twice the DPA compared to the newborns that received the plant oil–based formula. Studies of primates and human infants show that long-chain omega-3s are essential for proper brain, eye, and central nervous system development.

A balance of fats is important before birth too. Both the long-chain omega-3, DHA, and the long-chain omega-6, ARA, are vital for normal growth and development of the brain and vascular systems. Indeed, the placenta acts as a superpump for delivering them to the growing fetus, and if levels of either are deficient, serious problems in the central nervous system develop.

It is well established that a mother's stock of long-chain omega-3s, and especially DHA, progressively declines during pregnancy unless dietary intake compensates for diversion to the growing fetus. And it compromises them both when a mother's body depletes the stock of omega-3s she can deliver to her developing child. Low omega-3 levels in the last trimester of pregnancy have been associated with enhanced postpartum depression. Unfortunately, American women now have among the lowest breast milk DHA content in the world.

Low maternal intake of long-chain omega-3s during the second half of pregnancy appears to affect a child's IQ—perhaps for life. A randomized, double-blind study recruited Norwegian moms-to-be to consume 2 teaspoons a day of either omega-3-rich cod liver oil or omega-6-rich corn oil starting halfway through pregnancy to 3 months after giving birth. Cod liver oil is rich in both DHA and EPA, whereas the corn oil had 50 times more omega-6s than omega-3s. The breast milk of mothers who got cod liver oil contained 3 times more DHA and 10 times less ARA than the breast milk of mothers who got the corn oil. When the children from both groups were tested at 4 years of age, those whose mothers received cod liver oil had significantly higher mental processing scores than those whose mothers got corn oil.

Mental health and behavioral issues are another intriguing area of dietary omega-3 research. Increasing consumption through either diet or supplementation has been associated with reduced symptoms of depression and bipolar disorder, behavioral problems in children, and aggression and impulsive behavior among young men with no history of either. Violence among adult prisoners and repeated criminal offenses also declined with higher consumption of omega-3s. Of course, as Ancel Keys inadvertently showed us, correlation does not establish causality. But striking parallels between rising homicide rates and increased omega-6 consumption from 1961 to 2000 in Argentina, Australia, Canada, the United Kingdom, and the United States led researchers at the National Institutes of Health to suggest the possibility of serious population-scale behavioral effects of the modern dietary shift.

Several meta-analyses of the efficacy of omega-3s in treating mood disorders found promising potential for treating depression. For example, a study of older adults reported fewer depressive symptoms for those with lower ratios of omega-6 to omega-3 fats. Patients with major depression also had significantly higher cytokine production than those who were not clinically depressed. What helps curb inflammatory cytokines? Omega-3s.

Degenerative brain diseases that develop later in life also have been associated with low omega-3 consumption. One study involving over a thousand middle-aged participants found that lower consumption of marine omega-3s (fatty fish) significantly increased the risk of cognitive impairment. A similar study of a hundred older adults without cognitive impairment found that omega-3s supported the ability to reason abstractly and solve problems.

A well-controlled clinical trial found that omega-3 supplementation over the course of a year improved memory in patients with mild cognitive impairment. Another study that followed almost a thousand elderly men and women for just under a decade found that those with the highest blood plasma level of the long-chain omega-3 DHA had almost half the risk of dementia and Alzheimer's. And a study of more than a thousand older patients found that those with lower blood levels of long-chain omega-3s exhibited accelerated brain aging, with smaller brain volumes and lower scores on tests of visual memory and abstract thinking. On top of all that, magnetic resonance imaging of brain tissue from several thousand patients over the age of 65 found a 40 percent lower risk of blood supply obstructions for those with the highest blood levels of long-chain omega-3s. Considering these findings all together, it seems reasonable to suspect that dietary changes in the balance of omega-6s and omega-3s may have contributed to the alarming recent rise in dementia and Alzheimer's.

It's not just us who thinks this may be possible. Interest in dietary approaches to prevent or slow the progression of neurodegenerative diseases is growing due to the lack of other effective therapies. A 2019 review of clinical trials concluded that for early stages of disease an increased dietary intake of omega-3s offered a safe, well-tolerated, and potentially valuable treatment. In other words, higher omega-3 consumption may not be a silver bullet treatment, but it does appear to offer a low-risk way to help maintain mental capacity as one ages.

An interesting laboratory study shed light on how all of this may work. When a roundworm gene for converting omega-6s into omega-3s was introduced to a culture of human breast cancer cells, the ratio of omega-6 to omega-3 of the cell culture fell from 12:1 to less than 1:1—a drop roughly equivalent to reversing the modern increase in the Western diet relative to humanity's preagricultural diet. This change not only significantly reduced proinflammatory markers and inhibited the cancer cells in the culture from proliferating; it ultimately led to cancer cells dying. The researchers also found that while the cancer cell culture receiving the omega-3-converting gene died, a similar culture that did not receive the novel gene kept growing unchecked.

An ingenious experiment with transgenic mice provided a direct demonstration that omega-3s are causally linked to tumor prevention. Genetically engineered, omega-3-producing mice were crossed with another line of transgenic mice engineered for aggressive breast cancer. The experimenters then fed either an omega-6-rich diet or an omega-3-rich diet to the pups of the cross (those with the omega-3 and tumor-producing genes) and to a group of the tumor-fated mice that did not get the omega-3-producing gene. Both groups of mice with higher omega-3 levels—those that ate omega-3s and those genetically engineered to make their own—exhibited far fewer, smaller tumors over their lifetime. The researchers concluded that their experiment provided unequivocal evidence of a causal link between omega-3s and tumor prevention. Of course, mice are not women or men, but these findings prove interesting nonetheless.

All in all, there is little question that an omega-3-rich diet favors better health, and a diet swamped with omega-6s hampers formation of long-chain omega-3s. While substantial variability in clinical trials in part reflects the wide variety of influences on our health, as well as our genetic and biochemical individuality, a number of epidemiological studies report broad evidence for beneficial effects of higher omega-3 consumption in human populations.

A particularly interesting Greek study gave dozens of middle-aged men with high cholesterol levels but no evidence of heart disease three teaspoons a day of either omega-3-rich flaxseed oil or omega-6-rich safflower oil. Based on the composition of the oils, this made for a ratio of omega-6 to omega-3 of just over 1:1 for the participants given flaxseed oil and more than 10:1 for those receiving safflower oil. Here was a rough stand-in for the dietary switch to a modern Western diet. After several months there was no change in the group that received safflower oil. But the average blood pressure in the group receiving omega-3-rich flaxseed oil fell several times more than the average reduction in blood pressure reported in 20 randomized controlled trials of statin drugs typically prescribed to lower cholesterol and thereby reduce blood pressure. In other words, this study found that changing dietary fats worked as well as our modern go-to drugs.

A study of global patterns of disease provides further evidence for the health benefits of a lower dietary ratio of omega-6 to omega-3 fats, with the Japanese diet estimated to best meet dietary needs with a rough balance of omega-3s and omega-6s. The study concluded that omega-3 deficiency could indirectly account for roughly a fifth of all mortality among men and more than two-thirds of people with depression. Even if these conclusions are wildly overdrawn, the study reinforces the need for adequate levels of dietary omega-3s.

Reducing excessive inflammation is a major tool and goal of modern medicine. Nonsteroidal anti-inflammatory drugs, which inhibit an enzyme that normally activates omega-6-initiated inflammation, pull in global sales amounting to more than $10 billion annually. But there is another way to modulate inflammation—how we grow the crops and raise the animals that become part of the human diet. Just don't expect pharmaceutical companies to embrace the idea.

Perhaps we should consider the goal of increased omega-3 consumption or supplementation not so much to improve health but to help a body maintain it. So how do we get more omega-3s in our diet? Eat them. More leafy greens, more vegetables like broccoli, more nuts like walnuts, and to the extent possible more wild, cold-water fish. Of course, another way to reduce your dietary ratio of omega-6 to omega-3 is to eat fewer omega-6s, and a simple way to do that is to eat less processed food full of omega-6-rich seed oils. Switching over to grass-fed meat, dairy, and eggs instead of conventional varieties offers yet another way to achieve a better balance of fats.

How much of a difference would eating grass-fed dairy and meat actually make in a typical diet? An analysis that addresses this question stems from a study that modeled the amounts of omega-3s and omega-6s in a series of diet scenarios using values for Grassmilk and conventional milk for dairy products. The modeled food choices did not involve major dietary changes but simply replaced conventional dairy with Grassmilk dairy. The model's baseline scenario for the typical American diet assumed a ratio of omega-6 to omega-3 of just over 11:1. For diets with moderate to high dairy consumption and typical high-omega-6 sources for nondairy products (processed foods), switching to Grassmilk dairy products cut the overall dietary ratio to between 6:1 and 9:1. For scenarios that included both switching to Grassmilk dairy products and reducing nondairy omega-6 sources, the ratio fell to between 3:1 and 4:1, which is a whole lot better than the estimated 10:1 or 20:1 ratio typical of Western countries. Replacing conventional meat and eggs with 100 percent grass-fed would bring the balance of omega-6s and omega-3s even closer to 1:1.

How many human health problems like heart disease that we associate with red meat and dairy stem from how we feed cows? That's hard to say with the studies that have been carried out so far. But research documents underlying mechanisms, and epidemiological studies across a wide range of ailments point to reduced risks of a diet with a balance of omega-3s and omega-6s.

Still, it is naive to think that any particular dietary regime will benefit the health of everyone. Whether fats are good or bad for one's health depends on the specific types and amounts, their relative balance, and each person's unique biochemistry and genome. Nonetheless, this leads to some simple, well-grounded recommendations about dietary fats. While you may not want too much saturated fat in your diet, you really do want lots of CLA and a balance of omega-3 and omega-6 fats. So if you choose to eat animal products, consider sticking with modest amounts of 100 percent grass-fed meat, dairy, and eggs—and as much wild salmon or other cold-water fish as you can get, afford, or stomach. And pile your plate with omega-3-rich whole plant foods instead of omega-6-rich processed foods. It's not rocket science. It's a balancing act.

FILLING EMPTY PLANTS

The art of healing comes from nature, not the physician.

—PARACELSUS

The effect of milling on the taste and quality of bread snapped into sharp relief for us on a visit to France while writing this book. After an overnight flight to Paris we took a fast train southwest to Le Mans and arrived around noon. Weary and hungry, we dropped our things at our Airbnb and set off in search of some cheese and a baguette.

Within a few minutes we spied "Boulanger" on a shop façade, crossed the street, and browsed rod-shaped, round, and oval loaves big and small along with all kinds of pastry creations. After a quick stop at a grocery store for cheese, we were soon back at our Airbnb tearing into a light-gold baguette. It definitely had the requisite crunch: bits of crust went flying. But it was, alas, utterly devoid of flavor. The next day the remaining bit of baguette shattered like glass when we tried to break off a piece.

What made that French baguette particularly disappointing for us was that it literally paled in comparison to what we'd been buying at a bakery that had recently opened in our neighborhood. Rob, our hometown boulanger, made a whole-wheat baguette from einkorn, an ancient wheat variety. Nutty, satisfying, and seemingly incapable of going stale, Rob's baguettes were unusual but really, really good—unlike anything we'd had before. They simply blew away what we found in France.

A few months later our little bakery stopped making the delicious baguettes. Rob shrugged: "They just weren't selling." People wanted a French-style baguette, smooth and crunchy on the outside, pale and white on the inside. He couldn't afford to toss product, so he reverted to standard white-flour baguettes. This led to longer conversations with Rob about baking and wheat that eventually brought an unusually broad-thinking wheat geneticist to our attention.

The Bread Lab

Like us, Stephen Jones lives in the northwestern corner of the country. So a few weeks later, on a sunny day in late July, we drive an hour and a half north to the Washington State University Breadlab in the Skagit Valley, one of the state's premier agricultural areas. The large business park–like building looks odd, but we know we are in the right place when we spot the lanky wheat plants in pots flanking the front door. The real giveaway stands just inside, where a massive weather-beaten wooden contraption supporting a heavy stone grinder dominates the entryway. Little did we know this 1870s-era wheat mill was the crown jewel in the collection of a wheat-obsessed rebel.

Greeting us with the story of this curious mill, Jones ushers us into the building, looking every bit the farmer-professor hybrid in gray work pants and a blue plaid shirt that set off his dark salt-and-pepper hair and lean build. With an easy smile and a humility refreshing to find in the academic world, he is the fifth person to direct a century-long wheat breeding program at Washington State University. He approaches tweaking wheat genetics the old-fashioned way—through cross-pollinating different varieties, planting the resulting seeds, seeing what comes up from the ground, and doing it over and over again to select for favored traits.

Jones guides us down the hall and into a room that doubles as a wheat mill museum, all the while talking us through the history of milling. He shows off stone mills like the Greeks and Romans used, where a slab of rock rotates and grinds over a stationary one. The ancient designs seem awfully similar to the pioneer mill in the foyer. Part of the appeal of stone mills is that everything in the grain comes back in the flour. If you put in 100 pounds of wheat, you get 100 pounds of flour back out.

That's not what you get from modern steel roller mills designed to grind and separate the main parts of a wheat grain. To demonstrate, Jones pours some into the top of one of the mills scattered around the room. He turns the crank for a bit and opens the drawer at the bottom of the mill to reveal two piles segregated into separate trays of pale-white flour and darker bran. The visual contrast says it all. Modern milling turns wheat into snow-colored fluff. Holding up the brown tray, Jones points out, "Everything we want is in there—the fiber, iron, and zinc."

Today, bran and germ get cast off to feed livestock.* In so doing, we're taking the most nutritious part of a human staple and feeding it to animals that evolved to eat other things. The bran, germ, and other discarded parts of the seed coat are rich in fiber, minerals, and phytochemicals. A 2016 review concluded that wheat loses almost three-

* The dark outer layer of a wheat kernel is the bran. It accounts for less than 15 percent of the kernel weight but contains most of the fiber, vitamins, and mineral micronutrients. The plant embryo, the part known as the germ, makes up less than 3 percent of the total mass. It is rich in micronutrients, B vitamins, antioxidants, and fats. The pale, starchy interior is the endosperm. It makes up more than 80 percent of the total mass, consists mostly of carbohydrates and protein, and provides energy to sustain newly emerged plants until they can photosynthesize on their own.

quarters of its vitamins and minerals when milled into white flour. In contrast, whole-wheat flour retains virtually all of the original nutrients and phytochemicals.

Jones is unusual for a wheat breeder. He's as concerned about nutrition and flavor as yield, frustrated that crop breeders and nutritionists work independently. "Food science," he sighs, "is a field all about processing." Yet among the many things that vary between wheat varieties are the vitamins and minerals they take up and the phytochemicals and fats they make and store in their seeds.

Wheat was around long before people noticed and began eating it. Along with corn and rice, it descended from a common ancestor more than 50 million years ago. The modern wheat we know today comes from crossing three ancestral species (spelt, emmer wheat, and goat grass) to produce a grain with three sets of seven chromosomes. This makes for lots of genetic variability. All told, the wheat genome has some 16 billion DNA base pairs. Considering that the human genome has only 3 billion, this gives wheat breeders a lot to work with.

The genomic diversity of wheat reflects the great range of environments in which wheat naturally grows. Yet for thousands of years farmers have been selecting wheat mostly for a singular purpose—more grain. Breeders still do today. But not Jones. He looks beyond yield to tap into variation in other properties as well.

Jones takes issue with those who fetishize ancient, low-yielding wheat varieties as well as "the move-along-everything-is-fine" approach conventional breeders follow. He sees plenty of variation in the wheat genome from which to breed for both yield *and* nutrition. It's just that hardly anyone tries for both. Jones works the middle ground, seeking to increase diversity and enhance both flavor and nutrition without sacrificing yield. He's also concerned with another fundamental—baking performance. After all, that's what wheat is all about. Few wheat breeders, however, seek input from us eaters on what we want from a loaf of bread.

Relative to today's common wheat, one study found that ancient varieties had a third more to half more zinc, a third more iron, and up to a quarter more copper. Einkorn, one of the original varieties of wheat,* has a significantly higher antioxidant content, with up to ten times the carotenoids of modern bread wheat. Ancient varieties also have a lot more fiber. Those delicious Seattle baguettes really were a lot more nutritious than the tasteless French ones.

There is ample evidence that eating the whole of a cereal grain benefits our health. Epidemiological studies report that a fiber-rich, whole-grain diet helps lower cholesterol and blood pressure and may protect against heart disease. A 2000 review of previous studies concluded that diets with the highest intake of whole grains significantly reduced the risk of heart disease. Six years later, a dietary study involving several dozen men and women fed a controlled diet substantially reduced their blood pressure by replacing a fifth of their caloric intake with whole grains.

Whole-grain diets have also been found to help protect against type 2 diabetes. Decade-long studies of tens of thousands of American men and women found that those who consumed the highest proportions of whole grains had roughly a 40 percent lower risk of diabetes than those who ate the least. A similar study of over 4,000 Finnish men and women found a 35 percent lower risk of diabetes for those with the highest consumption of whole grains relative to those who consumed the least.

Eating whole grains also reduces the risk of some types of cancer. For example, a study involving more than 23,000 women in Iowa found that the risk of uterine cancer was more than a third lower for those with the highest whole-grain consumption relative to those with the lowest. An earlier study involving thousands of cancer patients in

* Einkorn, emmer, and spelt were the three earliest cultivated wheat varieties in the Fertile Crescent of the Middle East. Collectively they are known as ancient grains.

northern Italy found that high whole-grain consumption consistently and substantially reduced the risk of colon, breast, prostate, and bladder cancer. Researchers have attributed the protective effects of whole grains to their greater mineral, phytochemical, antioxidant, and fiber content relative to processed grains.

Yet for the past century, conventional wheat breeders almost exclusively focused on yield and shelf life—on maximizing volume and storage potential. While Jones thinks that regional breeding to increase diversity could jack up the mineral content to "make wheat with three times the iron," there's little point if processing just strips it back out. If we want more nutritious bread, pasta, and crackers, we need to rethink not only how we farm but how we breed, mill, and process the grains that power our bodies.

Wheat is a dynamic, living genetic library. Many wheat seeds still germinate after ten years, and nearly all will do so after three or four. So farmers and breeders can keep seeds viable for a while, but only for so long. You have to plant seeds while they remain viable to get new plants that produce more seeds. So those planted today govern the options for varieties to cross in the future. It's an ongoing, never-ending experiment guided by generations of farmers and breeders selecting for properties they like—over and over again through centuries of harvests.

Escorting us to a large room at the back of the building, Jones opens the door to an inner sanctum of sorts—his wheat seed repository. It dazzles. At least a thousand varieties of wheat line the walls in jars labeled by the region and year they were "discovered"— France 1838, China 1851, and Uzbekistan 1900. Closer to home is Mammoth Red 1904, out of Oklahoma. We do a three-sixty to scan the shelves and take in the natural kaleidoscope of jar upon jar of white, light tan, brick red, dark brown, and even blue-hued grains. But for all the diversity surrounding us, virtually none of the darker varieties are grown on the millions of acres of commercial wheat across the country.

A lot of things besides color also vary among wheat varieties, from the size and shape of kernels to other traits you won't know about until you grow the plants out. Some varieties grow tall. Some stay short. Others develop seeds earlier than average, or later, an adaptive response to variable frost timing. Some years it's good to be early; sometimes late proves better. Such differences affect crop performance, yield, and nutrition. This variability gives breeders a lot to work with.

Jones started breeding wheat for Washington State University in 1991, creating nine varieties for commercial growers. Since the program started in 1894, it has produced over 100 new varieties. One of his predecessors even provided the original seeds for Green Revolution wheat. Breeding for a shorter plant prevented its amber waves from crashing to the ground in a field. It also increased the proportion of biomass that was harvestable—the wheat we eat.

Crop breeding sounds deceptively simple. Try a variety of crosses, look for traits you like, keep those and throw away the others. Then keep doing it. "Breeding is not preservation," Jones explains. "It's moving forward." Life doesn't stand still. While most breeders habitually cross the best with the best, ending up with pretty similar varieties, he tries to introduce variability to increase genetic diversity. Now he's working to boost nutrition and baking performance in high-yielding grains.

Lunchtime arrives, and we head into one of the lab's kitchens. Jones cuts into a loaf of rye made from his grain. Rich and delicious, this whole-grain version bears little resemblance to typical mass-produced, caraway seed–flavored loaves. The Breadlab rye is moist with a nice tangy flavor tinged with sweetness. We sit on tall stools around a wood-topped table, enjoying the bread, while we talk with Jones and his graduate students.

Jones related the story of how celebrated New York chef Dan Barber noticed the difference in wheat grown on different sides of Washington State. Barber had teamed up with Jones to find a particularly delicious flour for his restaurants. At one point Jones ran out of wheat grown in rain-soaked Western Washington, so he shipped Barber some of the same variety grown in semiarid Eastern Washington. Barber called Jones to ask what happened to his wheat. It tasted different! Who knew that, like wine, wheat had terroir, a distinctive taste of place? This aspect of breeding—how a place and its environment affect flavor and nutrition—is another point of interest for Jones.

Merri, a reformed nutritionist who keeps a full drum set in her office, chimed in to note that wheat gets overlooked a lot of the time in regard to nutrients. She's interested in breeding with an eye to putting a thicker bran layer on the seed to increase fiber, minerals, and vitamins. She relates that her training as a dietitian never addressed how food is grown affects nutrition. "They teach that a potato is a potato. But it seems like a no-brainer that how it's grown will matter."

In her experience, hospitals and treatment centers mostly wanted to feed people cheap food. While Merri thought her clients would be better off growing their own food and cooking it themselves, she quickly learned that's not how the treatment industry works. Instead, she was supposed to teach "there's no such thing as good and bad food" and emphasize the need to eat anything in moderation. Merri thought that failing to prioritize fresh, whole foods as the foundation for good nutrition missed the mark. She quit after a new boss admonished her to stick with providing dietary advice on carbs, fats, and protein intake: "You need to teach them the basics of nutrition and not a passion of yours." She landed at the Breadlab.

Robin, a chef from a small town in Italy, started out working on blue wheat, looking at whether it had different antioxidant properties from the red and white varieties sold in global commodity markets. He didn't find big differences. So he shifted to working on a new perennial grain, Salish Blue, a cross between bread wheat and wild perennial wheatgrass that Jones developed over several decades as a grain and forage crop. The idea was to make a wheat-like grain that does not die each year, so farmers would not have to till and replant annually.

This would reduce erosion, save on diesel and herbicides, and build soil organic matter. As a commercial bonus, this blue-seeded hybrid is a natural cross and thus certifiably non-GMO. This new crop could improve soil health. But would it also be better for those who eat it?

Fields of Difference

Jones wants to show us where he grows his wheat, and so we follow behind his blue Chevy pickup and drive over to the Mt. Vernon extension center farm. We park and walk toward a fenced area and through a gate to find ourselves in a world of wheat, ribbon-like leaves fluttering and rustling in the wind. Eye to eye with the tops of plants in some patches, we reach down to touch the seed heads of plants in others. A single, thin-stalked plant looks awkward and gangly, but thousands standing together in a field help hold one another up.

Jones explains we are looking at a 20-acre area divvied up into a series of 4-by-12-foot plots, each containing a distinct wheat variety. The first few rows are where Jones tries out new crosses. Certain plots are staked with orange or blue pennants. Orange means a flawed cross, not to be repeated. Blue signals a plot with particularly desirable traits, like robust, unbroken stems or large seed heads, indicating a cross to be harvested by hand and saved for breeding. Each year Jones goes through and picks what looks best, has good characteristics, or has potential for useful variation. His goal is to capture the best of the variability that comes up in his fields.

We go deeper into golden-brown fields of conventional winter wheat in what he calls his advanced breeding nursery.* These stands are about four feet tall; almost all of it looks pretty good, with well-developed seed heads and no broken or bent stalks. These plants passed

* Winter wheat is planted and sprouts in the fall, goes dormant over the winter, and then grows again in the spring to be harvested in summer. Spring wheat is planted in the spring, grows over the summer, and is harvested in fall.

the trials in prior years and display the variability he wants to show us. Some stands have squat, densely packed, two-inch-tall seed heads that stand vertical on the plant. Others have less densely packed, flexible seed heads about five inches tall arching in an upside-down J shape. Eye-catching awns resembling cat whiskers extend from each seed head. Others stand nakedly awn-less. Some seed heads are light colored, others deep burgundy to brown.

Plots blur together as we move among them. But as Jones points out the kinds of subtle differences his eyes see, we start to see them too—a small difference in the length of an awn, a seed head that sits at a particular angle, a slight color variation, or a fragile stalk that will probably break in high winds. There's a lot to notice, and with tens of thousands of different varieties of wheat in the world, we wonder how someone could master all this. Through a quick smile, Jones says, "It helps if you've looked at wheat every day since 1977." He can pick out a wheat plant and envision how it might do in a future cross. Persistence and an eye for detail are what make a great wheat breeder.

Modern commercial stands grow short and uniform in size. In contrast, breeding for diversity gives you variability even in a field of the same wheat variety. This genetic diversity is particularly important for resilience under limiting conditions such as drought or lack of fertilizer. It's also important for nutrition and for finding traits good for organic and low-input systems, an increasing focus of Jones's efforts.

As we talk, he walks us over to his spring wheat stands. Some are golden, some still green, and some brown and already setting seed. Yet they were all planted on the same day. "All the traits we've looked at in the field today are from a single gene," he says. Even the difference between winter and spring wheat lies in a single gene. Jones estimates there are many thousands of distinct genetic variations among the wheat plants in his fields. Looking ahead, he thinks farmers will need to draw on this variability to weather climate change and other future challenges. Jones gestures across the plots: "We've got all the diversity we need out there."

When his lab releases a new wheat variety to farmers, they put a name on it. Their most recent release—Salish Blue, the variety that chef Robin mentioned to us earlier in the day—is not just a new variety but a new species. Salish is the name for the Indigenous people in the region, and blue reflects the grain color. A winter perennial, it expects to be around for more than a year, so it invests in building belowground biomass in its root system. Because this translates into more soil organic matter, Jones wants to use Salish Blue in crop rotations to help regenerative farmers improve the health of their soil.

Jones also breeds wheat for organic farms. It's little appreciated beyond wheat breeders that most modern varieties were never bred for growing in soils with high levels of organic matter and robust populations of beneficial soil organisms—like on a regenerative farm. Though his organic stands look pretty good, the soil looks much like the crappy, low-organic-matter soils of the nearby conventional wheat plots. A bit sheepishly, he notes that until recently "recreational tillage" occurred on most of the research farm. While he says the soil in the organic fields is improving after good rotations over eight years, the farm is managed in a way that tillage still occurs frequently enough that soil organic matter remains below 2 percent. As with David Johnson's experience in New Mexico, this experimental farm long emphasized crop yields, with no apparent interest in building or investigating soil health. It seems researchers wanted to stay relevant to conventional farmers trying to grow crops in typically poor soil. Now that's starting to change.

Most conventional farmers overfertilize wheat to boost protein content, as plants shunt excess nitrogen to their seeds, making for higher-protein grain. So breeders generally select for plants that thrive in nitrogen-rich environments. Jones says, "The plants become addicted to nitrogen instead of foraging for themselves. We know that. There is no doubt about that."

But breeding for high protein levels creates problems too. In wheat, high protein means high gluten. So more nitrogen fertilizer yields wheat with higher gluten content. The gluten isn't there for us; it's a storage protein for the seed. A high-gluten flour is also easier for bakers to work with, a welcome feature in facilities that mass-produce bread. Jones says it's no accident that the composition and strength of gluten in wheat rose dramatically in the past half century. Breeders selected for it.

Growing wheat bred for conventional methods in an organic setting generally doesn't work out well. This is the case for other crops too. Plants can prove as confounded as a herd of cattle moved to an unfamiliar landscape or land ill-suited for grazing.

Jones maintains that "breeding is designing, whether classical breeding, GMOs, CRISPR or whatever." You have to know what farming system you are designing for to know what variations to select for propagating. Conventional crop varieties were developed to be fed a lot of nitrogen, so they do well under those conditions. But breeders can select for other things too, like variations more or less suited to different levels of organic matter—for either degraded or healthy soils. Jones foresees no problem with breeding for high-yielding crops with low need for nitrogen fertilizer in healthy soil.

So that's one of the things he's now trying to do. The conventional varieties farmers grow were not bred to perform well in organic systems. So they don't. Plants bred for high nitrogen environments don't thrive in low-input systems and vice versa.

He sees demand as the limiting factor for breeding programs that select for nutrition. Most frustrating to Jones along these lines is consumer perception that whole wheat means worse bread, the problem our neighborhood baker ran into. "Imagine if we put as much energy into whole wheat as we have for white bread over the past century. There's not a pastry made that can't be made with whole wheat." He's confident that turning this situation around through creating demand for food grown in healthy soil would get breeders working on growing and developing high-yielding, nutrient-dense strains well adapted to regenerative farming methods.

Jones insists it's not really a question of yield versus nutrition. His highest yields of spring wheat have been in organic fields planted after alfalfa. But to get high yields under an organic system, you do have to breed for them.

He thinks taste and flavor can help drive change back toward whole-grain flours. Good chefs can tell the difference between whole-wheat croissants made from grain grown in different locations. This isn't the case with breads made of white flour, as the parts removed through milling—the bran and germ—are the parts of the grain loaded with flavor-suffusing nutrients, fats, and phytochemicals. Cast them aside and flavor takes a big hit. Jones harbors no doubt that the condition of the soil will affect the flavor of whole wheat. While he can't recall a baker ever asking about breeding for flavor in the 20 years he bred commodity wheat, he now fields such inquiries all the time.

In general, though, he says people working in wheat breeding don't think much about minerals, phytochemicals, or flavor. But that doesn't mean they can't. After all, there's a simple way to breed for more iron or zinc: look for varieties with higher levels and use them in wheat breeding programs. Though our modern cereal crops inherited the ability to interact with mycorrhizal fungi that can acquire and transport zinc to plants, so far breeders have neglected to select for varieties that do so.

We finally get around to the original question we came to ask Jones: did he know if ancient grains and regeneratively farmed ones were nutritionally superior to conventional? Just as for flavor, in all his years of wheat breeding no one had ever asked him about breeding for nutrition.

And what about soil health? What does healthy soil mean for micronutrient levels? He didn't know, but he offered to get us some samples of wheat grown at the university experimental plots and from several other farms in the region. A few weeks later the wheat samples arrived: Salish Blue and a conventional wheat variety known as 1109 grown under conventional and organic methods.

When we got the results back from the lab, the differences in vitamins and minerals were striking. Compared to the 1109 variety grown organically, the samples from Jones's organic Salish Blue plots had about 50 percent more vitamins K and B_1, six times the vitamin B_6, and more than ten times the vitamin B_5, as well as three-quarters more calcium and manganese, 43 percent more zinc, 60 percent more iron, and 90 percent more copper. The same comparison for the two wheat types grown conventionally revealed that Salish Blue had 29 percent more vitamin K, about twice as much vitamin B_1, a third more vitamin B_3, eleven times more vitamin B_5, and three times more vitamin B_6, as well as 18 percent more zinc, a third more manganese, almost half more calcium, and two-thirds more copper.

Averaged over all four farms regardless of growing method, Salish Blue had 29 percent more vitamin K; about 20 percent more of vitamins B_1 and B_3; more than twice as much of vitamins B_2, B_5, and B_6; 19 percent more zinc; a third more manganese; half more calcium; and two-thirds more copper. But 1109 is no nutritional slouch to begin with. Consider that while the USDA nutritional database lists whole-grain wheat flour as having 32 parts per million (ppm) of zinc, the 1109 samples averaged 34 ppm. Salish Blue averaged 28 percent higher at 41 ppm. In other words, however you look at it, Salish Blue tested out as more nutrient dense than conventional wheat: it's better for us as well as the land.

What's Not in Your Cereal

Jones's approach to wheat breeding departs radically from the mindset of breeders in the aftermath of the Second World War when global population numbers began rising rapidly. The specter of mass starvation and regional food shortages meant that issues of nutritional quality did not really factor into developing the high-yielding cereals at the heart of the Green Revolution. High yields were the goal. Cereals produced the most calories, so they received far more research support than vegetables and lower-yielding crops.

As yields of wheat, rice, and corn more than doubled, their price fell relative to other foods. So more people ate more of them. Many low-income people around the world now depend primarily on a simple diet of staple grains. In rural India, for example, cereals contribute almost three-quarters of the caloric intake for those at the bottom of the economic ladder.

A dietary shift to heavier reliance on cereal grains coincided with breeding efforts intently focused on yield rather than nutrition. As a result, iron and zinc concentrations declined in new varieties. And in the case of rice, carotenoids were also lost. They lent yellowish colors to rice, a feature not in line with the whiteness breeders sought. This preference for lighter color bred precursors to vitamin A out of rice. Decades later another scientific discipline stepped forward to genetically engineer the ability to produce beta-carotene (a precursor to vitamin A) into modern varieties to create "golden rice."

As the human diet shifted toward greater consumption of the new, high-yielding cereals and the amount of edible biomass in grain crops increased (mostly due to the increased size of individual grains), micronutrients like zinc and iron were spread thinner, diluting mineral density on a per-seed basis. Widespread micronutrient malnutrition followed as humanity harvested yield at the expense of health.

This mattered. A 2012 review attributed the rise in human micronutrient deficiencies to replacing nutrient-dense pulses and other noncereal crops with micronutrient-poor cereals to feed the burgeoning human population. On top of this dietary shift, refining and processing of cereal grains further exacerbated micronutrient deficiencies. Zinc deficiency in particular grew into a critical, widespread side effect of the Green Revolution. As both iron and zinc are critical micronutrients, the effects of breeding and cultivation practices on their levels in cereal grains are especially concerning.

Globally, nutritional deficiencies now cause two-thirds of all childhood deaths, with low intake of vitamin A, iron, and zinc at the top of the list. In particular, vitamin A deficiency affects about 250 million children worldwide, causing blindness in an unlucky quarter million every year.

While micronutrient deficiencies vary both geographically and socioeconomically, the diet of more than two-thirds of the world's population is deficient in at least one mineral. In West Africa, for example, half the women of reproductive age are deficient in iron. And a 1992 study of village-dwelling toddlers found that between a third and half in Egypt and Mexico did not get enough iron. More than half of those in Kenya had inadequate dietary zinc. More recently, a 2007 study based on a national nutritional survey in China found that almost a quarter of the country's children suffered from iron deficiency and over half were deficient in zinc. And while more than 70 percent of children in Cambodia lack sufficient iron and zinc, more than 10 percent of Americans don't get enough zinc.

It is no coincidence that zinc deficiency is the most common global micronutrient deficiency. Part of the problem is that zinc is an uncommon element that's not very soluble and tends to bind with clays and organic matter. So what little there is in soil is not very mobile, making microbial activity critically important for making zinc available to plants. This helps explain why phosphorus fertilizers tend to decrease zinc uptake by crops: they reduce fungal symbioses. Recognition of the importance of maintaining an adequate dietary supply of mineral micronutrients like zinc shines new light on the role of farming practices in human health.

First diagnosed in people in the early 1960s, zinc deficiency remained largely ignored until the early twenty-first century when it was recognized as affecting a third of the world's population. While zinc is important on its own, it also helps our bodies absorb the iron in our diet. And about a quarter of the world's population and almost half of preschool children are iron deficient.

Our bodies require iron to make many proteins and enzymes in addition to the hemoglobin that carries oxygen in our blood. Yet iron absorption in humans is complicated. Animal foods, especially red meat, usually contain iron at sufficient levels for human health, whereas plant foods typically have lower levels. In addition, the iron in red meat occurs in an easily absorbable form. In contrast, iron uptake from plant foods also depends on what else comes in a meal, particularly the amount of A, B, and C vitamins in whole grains and dairy. Iron deficiency was another unexpected side effect of Green Revolution crops, one that led to a rise in the prevalence of anemia among populations that became most dependent on them.

Still, the new, high-yielding grains rapidly expanded at the expense of other crops. Between 1961 and 2013, the global land area planted in high-yielding rice, wheat, and corn increased from essentially nothing to almost 80 percent of all grains. Those with higher micronutrient content—like barley, oats, rye, millet, and sorghum—dropped to less than a fifth of all grains planted. In India, between 2004 and 2012, consumption of Green Revolution rice and wheat doubled while consumption of more nutritious grains fell by half. Yet millet has nearly four times the iron in rice. And oats have nearly four times the zinc of wheat. At the same time, the iron and zinc content of wheat and rice declined such that in the half century from 1961 to 2011, the calories one needed to eat to ingest the recommended daily amount of iron rose by almost a quarter.

The bottom line? By growing an abundance of less-nutritious grains, many countries now face two distinct nutritional problems. Part of the population suffers from inadequate caloric intake and micronutrient deficiencies. Another segment suffers from obesity and diet-related chronic illnesses. While impacts differ between wealthier and poorer nations, both ends of the economic spectrum face serious problems.

While ancient wheat varieties produce lower yields than modern varieties under conventional cultivation, they tend to produce better yields on marginal lands and under low-input and organic farming systems. Screening of high-yielding varieties of both modern and ancient grains found that high yields do not necessarily exclude high antioxidant content (carotenoids and vitamin E). It's not a trade-off. We just haven't been breeding for high vitamin and phytochemical levels. But if we did, it could enhance the nutritional value of wheat, leading some to suggest that breeding for nutrient density could potentially double the micronutrient availability in major food crops.

Naturally, soils developed from different underlying rocks have different amounts of available mineral elements. Zinc, in particular, is relatively rare geologically. Most rocks contain very little. And if zinc isn't in the soil, it's not getting into crops. Soils low on zinc can lead to deficiencies in crops, livestock, and people. Globally, around half of the agricultural soils currently used to grow wheat are considered low in bioavailable zinc. In contrast, iron is deficient in less than 5 percent of the world's soils. But while it is abundant, plant uptake isn't guaranteed, which means animals and people can also become iron deficient. Indeed, despite rocks and soils being rich in iron, several billion people lack enough. It's a bit like nitrogen is to plants. We're surrounded by it, but our bodies have a hard time getting at it.

There is substantial genetic variability in plant genomes that relates to micronutrient uptake and yet doesn't affect yields. This means there is a lot of potential to increase the nutrient density of crops using traditional plant breeding. For example, we could improve the micronutrient density of crops through breeding to select for greater uptake of minerals like zinc through the ability to forge microbial alliances. Relying less on chemical fertilizers and more on biologically driven cycling would enhance the micronutrient content of the food that makes it onto our tables and into our bodies. We want to limit shortages of critical micronutrients to occasional events rather than experiencing them as chronic conditions.

The ability to form mutualistic partnerships with fungi and bacteria are heritable traits that could be selected for in breeding programs. But we've mostly done the opposite in terms of potential mycorrhizal and rhizobial partners, creating newer cultivars that are either less able to form mutualistic relationships or that fail to sanction microbial freeloaders that take exudates but provide little in return.

Crop varieties bred for high-input conventional agriculture lack traits needed to support plant health and performance in organic and low-input systems. This helps explain why conventional crop varieties tend to fare poorly when grown organically and why organic varieties typically underproduce when grown conventionally. It also suggests that breeding programs focused on performance under regenerative practices could improve nutritional characteristics as well as yields.

Refilling Food

Farming practices also contribute to micronutrient malnutrition. As we've seen, fertilizers and herbicides affect the mineral uptake of grains, as do the effects of tillage on mycorrhizal fungi. Low mineral density of crops can be increased either directly through micronutrient fertilization or growing crops in ways that enhance mineral uptake and build soil fertility and organic matter. In the former case, fertilization with mineral micronutrients typically involves adding zinc and copper to soil. This approach could prove essential for soils in which certain micronutrient levels are naturally low, although their actual delivery to a plant depends on microbial partners. Sometimes adding micronutrients like iron can be achieved through spraying it on a plant's leaves.

Spoon-feeding crops a lot of nitrogen and phosphorus undercuts fungal partnerships. So when modern crops lost their mycorrhizal partners, it came with a hidden nutritional price. Side-by-side comparisons of modern cultivars and older, traditional varieties of grains and vegetables report reduced nutrient density in wheat, maize, rice, broccoli, cabbage, and lettuce. This reduction is thought to result from selective breeding for high yield without regard for the effect on micronutrient and phytochemical composition.

Most plants have two ways to take up phosphorus and other minerals. There is the direct pathway of simply taking up what's dissolved in the water that roots pull up from the soil. This is what conventional soil tests measure—the amount of plant-available, soluble phosphorus. The second pathway runs through fungal partnerships. These separate pathways relate to different sets of plant genes. Particular ones upregulate exudate production when phosphorus is scarce and downregulate it when phosphorus is abundant. This means that crop breeding could select for the ability to partner with mycorrhizal fungi to potentially increase both yields and nutrient density.

Finding the right fungal partners for a crop is critical because different plants form symbiotic relationships with different fungi. For most plants, if they lack fungal partners, it's not for want of trying. Such failures typically reflect degradation, sterilization, or fumigation of the soil, leaving no partners to recruit. Recall that frequent use of phosphorus fertilizers, low soil organic matter, and regular tillage reduce fungal colonization. Conventional farming strikes out on all three.

This isn't exactly news. It has been known since the 1960s that phosphorus fertilizers depressed zinc uptake in corn. Back in the 1970s experiments at Pennsylvania State University showed that while phosphorus fertilizers suppressed mycorrhizal activity, inoculation with mycorrhizal fungi increased uptake of phosphorus, copper, and zinc. Soybeans with mycorrhizal associations took up twice the phosphorus from the same soil. Corn took up more too. Conversely, applying phosphorus fertilizers to inoculated corn and soybeans decreased zinc and copper uptake. All told, crops that partner with mycorrhizal fungi typically took up at least twice the copper and zinc. The researchers concluded that phosphorus fertilizers reduced these minerals in crops through both the dilution effect of growing larger plants and suppressing mycorrhizal uptake.

So far, the ability of fungal hyphae to take up and deliver mineral nutrients to plants has been firmly established for phosphorus, copper, iron, and zinc. Fungi can deliver up to 80 percent of the phosphorus, more than half the copper, and more than a quarter of the zinc that a plant takes up. They also help plants access nitrogen in organic matter. Numerous studies have shown that mycorrhizal fungi and rhizobial bacteria can increase phytochemical production, especially phenols and flavonoids, in their plant hosts. Mycorrhizal colonization also increases levels of vitamin C, terpenes, and antioxidant carotenoids (like vitamin A precursors), alkaloids (like quinine or caffeine), and cancer-fighting, sulfur-containing compounds (like sulforaphane).

The effects of mycorrhizal fungi on crop uptake of zinc are particularly well established. A 2014 meta-analysis of 104 studies that reported on 263 field trials found that across a range of crops, mycorrhizae increased zinc concentrations by up to almost a third. But getting zinc into biological circulation is just the first step: keeping it circulating is as important. Mycorrhizal fungi help with both.

Conversely, high applications of phosphorus fertilizers reduce zinc uptake, in part because the practice inhibits mycorrhizal fungi from colonizing roots. A 2008 study from the University of Western Australia found that phosphorus fertilizers decreased the zinc concentration of wheat grains by more than a third. Likewise, a 2012 study from China Agricultural University showed that while applications of phosphorus fertilizer increased grain yield, it as much as halved its zinc content. Yet soil micronutrient concentrations did not change over the course of the experiments. So the difference lay not in a lack of zinc in the soil but in zinc not getting into the crops under high phosphorus fertilization. It's a familiar pattern—growth and yield at the expense of crop and soil health.

Recharging Staples

Growing two or more crops together is common in parts of Asia, Africa, and Latin America and can increase mineral density. For example, a 2008 review found that intercropping of peanuts and corn could increase the iron content of the peanuts by almost half. In some cases, intercropping wheat and chickpeas increased the iron content of both crops by more than a fifth and almost tripled the zinc content of the chickpeas. Along these lines, a 2016 review of intercropping on mineral uptake concluded that cereals and legumes mutually benefit through enhanced production of root exudates that increase the solubility of phosphorus, iron, and zinc in the soil and thus their availability for plant uptake. Such increases could go a long way toward addressing mineral micronutrient deficiencies.

Another way to increase the nutritional quality of wheat is to increase soil organic matter, which generally makes micronutrients more available to plants. A 2018 study of farms in Ethiopia found that along with yields, the protein and zinc content of wheat grains more closely tracked levels of soil organic matter than mineral fertilizer applications. Soil organic matter was the strongest predictor for zinc content in wheat, leading the researchers to conclude that increasing soil organic matter could increase crop nutrient content enough to improve human health. Along these lines a study of zinc levels in wheat grown in central India found that organic wheat that produced comparable yields to conventionally grown wheat had almost 20 percent more zinc. High yields don't necessarily come at the cost of nutrient density.

Can adopting farming practices that build healthy, fertile soils rebuild the nutrient density of other grains as well? Underappreciated evidence for how changing farming practices can increase the nutrient density of rice shows that cultivating beneficial soil life can enhance another dietary staple.

Half the people in the world eat rice. Most is grown on small farms where large clumps of seedlings are transplanted densely into flooded fields (rice paddies). A century ago, nitrogen-fixing cyanobacteria provided the primary source of nitrogen for rice cultivation. But applying high amounts of inorganic nitrogen fertilizer proved detrimental to cyanobacteria. With the introduction of fertilizer-dependent Green Revolution varieties, farmers found that the more nitrogen they added to their fields, the more they needed to add just to maintain production levels.

In the early 1980s a French Jesuit priest, Henri de Laulanié, developed the system of rice intensification (SRI) after working for two decades with smallholding farmers to increase rice harvests in Madagascar. The method they developed involved planting fewer seedlings earlier in their development with no continuous flooding of fields and using composted organic matter rather than chemical fertilizers. The rice plants grown with the new system were healthier, disease-resistant plants with larger root systems that produced up to twice as much harvestable rice.

What was their secret? SRI methods cultivated soil life and nurtured soil health. Healthy, productive soil produced healthier, more productive plants. With their larger, deeper root systems SRI plants put out more than twice the exudates of conventionally grown rice and nurtured beneficial microbes in the rhizosphere. Under this system the crop benefits from microbial symbioses and exhibits greater resistance to storm damage and droughts.

In the early 2000s this system began spreading to India, Asia, Africa, and Latin America. Why was it catching on? The only purchase needed to use SRI methods is an inexpensive mechanical push-weeder. It's a twofer tool, aerating the topsoil around plants as it buries weeds in the soil, turning them into green manure. SRI methods raised yields by 20 to 100 percent or more, while reducing expenditures for seeds, fertilizers, and pesticides by more than half, a combination that could double a farmer's net income. Initially, it took more labor while learning the new methods, but for most rice farmers it then became labor-saving. More challenging, however, was that it required thinking differently about the soil-plant partnership—seeing microbial life as allies to cultivate.

By 2004, reports of remarkable yield increases for SRI rice came to the attention of mainstream researchers at the USDA and the International Rice Research Institute—the driving force behind Green Revolution rice. Several were extremely skeptical and penned high-profile critiques in trade publications and academic journals challenging the validity of reported yield increases that were well beyond what conventional methods delivered. They went so far as to accuse SRI researchers of practicing pseudoscience and wasting valuable scientific resources on frivolous pursuits.

But less than half a decade later, in 2011, an Indian farmer used SRI methods to set a new world record for rice yield. By then most of the farmers in his district also used the new methods. Reports from India, China, and Africa began showing that SRI rice really did increase average yields, generally by at least a quarter, with some farmers reporting far greater increases.

In 2017, an editorial in the journal *Nature Plants* pointed to higher rice yields using SRI methods as a key development for abolishing world hunger. Now more than 20 million farmers in over 60 countries are growing rice using some or all of the SRI methods. It seems that the strategy based on building soil health that conventionally minded academics initially argued sounded too good to be true is rapidly catching on among farmers. Why? It works.

Field trials in 2020 at a leading agricultural research institution in India found that SRI methods consistently and substantially increased both yields and the mineral micronutrient content of rice grains and

straw, as much as doubling the levels of iron, zinc, copper, and manganese in rice grains. A previous 2017 Indian study likewise found that rice grown using SRI methods produced grains with about a third more manganese and zinc, more than a quarter more iron, and almost two-thirds more copper. For populations relying on rice for their caloric intake these are significant increases.

Combining the practices behind SRI rice with bacterial or fungal inoculants can further boost both yields and micronutrient levels. A 2016 study by microbiologists at the Indian Agricultural Research Institute found that SRI rice had substantially more iron, zinc, copper, and manganese across a wide range of fertilization regimes. In trials with no fertilizer supplements, the micronutrient content of SRI rice came in at about twice that of conventional rice. The researchers also found that inoculation with either cyanobacteria or *Trichoderma* fungi further increased micronutrient levels, with increases of more than 40 percent in iron and zinc. Similarly, a 2019 study in Nepal showed that inoculation of SRI rice with *Trichoderma* increased yields by almost a third, with an even greater boost for heirloom varieties compared to so-called improved varieties. A key advantage of algal and fungal inoculants is that they can be prepared, managed, and applied by farmers with little expense, producing immediate financial benefits through increased yields.

SRI methods also can be combined with no-till farming of other crops, bringing the diversity of crop rotation into rice production. Indeed, adoption of conservation agriculture practices in India and Mexico has been growing rapidly for corn, wheat, and rice, humanity's three staple grains. When new crop varieties were introduced in the 1960s, farmers began developing combined wheat-rice systems across South Asia. Directly planting no-till wheat into nonflooded rice stubble began in the 1980s and has greatly reduced fuel costs and greenhouse-gas emissions while controlling weeds and increasing beneficial insect activity. Perhaps most important for farmers is that the

new system produces higher yields at lower cost, a recipe for improved farm profitability—and wider adoption.

It appears that healthy, fertile soil can help offset adverse health effects of micronutrient deficiencies that can lead to self-reinforcing links between poor soil, poor health, and poverty. All this points to the benefits of a diverse diet grown on healthy soils. Once again, what's good for the land is better for us too.

GROWING MEDICINE

Leave your drugs in the chemist's pot if you
can heal your patient with food.

—HIPPOCRATES

In October 2018, two hundred medical and agricultural researchers met near Baltimore to explore connections between soil health and human health. Not long after the meeting began, diverging views about soil emerged. Those from the public health world saw soils primarily as sources of toxins, pathogens, and disease. One by one, they cast soil as a potential threat to human health, focusing on its role in harboring pesticides, lead, and heavy metals in urban environments. While the idea that diet matters to health would have surprised no one at the meeting, it was left to a lone physician and those of us from the soil side of the meeting to bring up the potential for healthy soil to support human health through the nutritional quality of food. To many at the meeting this appeared to be a novel idea.

Yet such connections came to light back in the 1930s. Much like Howard and Balfour, Ohio dentist Westin Price investigated linkages between soil, diet, and human health. He had seen the effects of a sugar-rich diet on American teeth. Price suspected modern dietary change as the culprit underlying dental problems running rampant across the westernized world. Supported by a thriving Depression-era practice in Cleveland, he traveled the world to compare the health of populations uninitiated to a Western diet to those adopting the new way of eating. And the more places and people Price visited, the more he noticed it was not just oral health that seemed to vary with diet. Public health in general deteriorated when populations switched from their traditional fare to a modern Western diet.

Price considered it well known among dentists that "savages" living under "primitive" conditions had surprisingly excellent teeth compared to the generally wretched teeth of "civilized" populations. He thought it curious that his profession concentrated on treating poor dental health without looking into why Indigenous peoples lacking their services had such great teeth.

Over the course of a decade, Price traveled to remote areas in 14 countries on 5 continents in steamships, early planes, and dugout canoes, as well as on foot. He found that people in these regions not only had sound teeth and excellent oral health; they enjoyed good health in general.

He visited Swiss valleys isolated high in the Alps, Native American villages from Alaska to Florida, African tribes, Indigenous Australians, Peruvian and Amazonian Indians, Maori villages in New Zealand, and remote South Sea, Malaysian, and Scottish islands. In each location he compared the health and diet of isolated and recently westernized groups. While he was mostly interested in disorders of the teeth and mouth, he also investigated overall health and tuberculosis in particular.

In each location he analyzed the composition of the local diet, learning what people ate and whether they consumed raw, cooked, or processed foods. He also conducted chemical analyses to determine the mineral and vitamin content of native and westernized diets in each region.

Everywhere he went, local populations were increasingly switching to diets rich in white flour and sugar, along with canned or skim milk and highly refined (saturated) vegetable and seed oils. He found that adoption of a highly processed diet resulted in not only an explosion of previously unknown tooth decay but dental deformities, birth defects, and increased susceptibility to infectious and chronic diseases. In each region, people who still ate a "primitive" traditional diet were healthier overall and enjoyed well-formed, cavity-free teeth.

Yet Price found quite a range of foods in traditional diets. There was no single ancestral diet shared by peoples around the world. Some subsisted primarily on seafood, while others mostly ate wild game or domesticated animals and dairy. Still others thrived on fruits and vegetables. Certain cultures tended to eat their food raw; others cooked almost everything. The common elements among traditional diets, Price realized, were the one thing they lacked—highly refined, processed food—and the other thing they all had—some form of meat or dairy, even if just modest amounts.

Around the world, the nutritional difference between traditional and modern diets proved striking. "Primitive" diets provided four times more water-soluble vitamins and key minerals—and more than ten times the fat-soluble vitamins. Price saw a repeated pattern of dramatic declines in health when native peoples switched from a traditional, nutrient-dense diet to a Western-style diet rich in sugar and low in fiber, vitamins, and minerals.

He interpreted this globally consistent pattern as a strong case for the health-promoting effects of a nutrient-dense diet of fresh, whole foods—and for restoring soil health through farming practices that relied on fertility from biological processes rather than chemical inputs. He considered the quality of soil foundational for human health due to its influence on the vitamin and mineral content of the plants that grew from it and the animals that then grazed upon them.

He drew support from veterinarians who saw that healthy forages set the stage for healthy livestock. These colleagues were not outliers of their profession, and Price found that their experience and views on the effects of soil degradation on animal health could fill a small library.

In his own studies he too had found wide variation in the mineral content of different livestock forages, with calcium and phosphorus content varying as much as 10- to 60-fold, respectively. Typical forage at the low end of the ranges he found reported in government studies would not provide an adequate mineral uptake for a cow. And whatever shortages cows experienced carried through to meat, milk, and cheese. In contrast, values at the high end would prevent deficiencies. Cattle grazing on young, rapidly growing wheat or rye grass rather than eating grain-based concentrates, as was then starting to become common, stayed healthier and in better physical condition. Their offspring fared better too, with calves growing more rapidly and exhibiting "much higher resistance to disease."

Price also thought that differences in mineral levels among forages would pass through to people who consumed meat and dairy products. To investigate this idea he compared monthly records of the vitamin content in butter and cream to monthly records of mortality from heart disease and pneumonia for more than a dozen regions across the United States and Canada. In Price's era people mostly ate food from local and regional sources, and across this broad geography he consistently found that fewer deaths occurred in spring and fall months when the vitamin content of dairy products typically peaked. He pointed to a study from Toronto reporting occurrences of a variety of childhood diseases that tracked with seasonal variations in the vitamin content of local dairy products. Bearing in mind that such correlations can prove spurious—for example, the way murder rates and ice cream consumption both spike in summer—they nonetheless hint at the plausibility of connections Price drew from his travels.

By today's standards Price also held some rather odious views, expressed in derogatory terms for Indigenous peoples and opinions that danced around eugenics. Despite his shortcomings, Price's studies of nutrition and health remain relevant today. He understood that a diet of unprocessed, whole, fresh foods provides a solid foundation for good health, and he endorsed the role of mycorrhizal fungi in getting minerals and vitamins into food. Price saw that food quality reflected land quality. Lessons from his study of Indigenous diets take on new significance today given the dramatic rise in chronic diseases over the last half of the twentieth century that shows no sign of abating. Price saw how cheap, industrialized food heralded expensive medicine, noting with alarm how sickness in the United States "cost nearly half as much as food." Indeed, Americans now spend more on health care than on food.

Price's pioneering studies raised compelling concerns over the nutritional consequences of intensive food processing and now-conventional farming practices. He went on to write *Nutrition and Physical Degeneration*, documenting the differences in the health of people in various places he had studied. Yet like the ideas in Lionel Picton's book, Price's were summarily dismissed in conventional circles. A review of his book in the *Journal of Pediatrics* called it important and thought-provoking insofar as it related observed differences in traditional, non-Western diets to differences in dental and physical health. The reviewer, however, focused on Price going too far in ascribing differences in character and morals to dietary differences. It didn't help Price's case that it was becoming more profitable to treat chronic illnesses than to cure them or prevent them in the first place. His prescription was out of phase with the modern push for doctors to treat the practice of medicine as a standard profit-focused business.

Farming for Health

Since Price's day, relatively few studies have focused on linkages between farming practices and human health. Two European studies examined differences in susceptibility to allergies in children who attended different types of schools. One of the studies, of 675 Swedish children, found that those attending Steiner schools* were less susceptible to allergies compared to students who attended public schools. A similar comparison of Steiner schools and public schools involving almost 15,000 children across Austria, Germany, Sweden, Switzerland, and the Netherlands also found a lower incidence of food allergies, as well as asthma. The organic, vegetable-rich diet of Steiner schoolkids and their families was cited as a primary contributing factor.

How food is grown affects adults too. A 1996 study published in *The Lancet* found that relative to the general public, sperm counts were 43 percent higher among Danish men who consumed at least a quarter of their diet as organic produce. A subsequent study of Danish farmers found that those whose fruit and vegetable consumption was more than half organic had the highest percentage of normal sperm. Farmers who ate only conventional produce had the lowest. Perhaps those bitter taste receptors in the testes are there for good reason.

A Danish dietary intervention study employing a rigorous, double-blind, randomized, crossover design compared the intake and excretion of five flavonoids for volunteers who consumed a prescribed diet of either conventional or organic foods. Each subject was randomly selected to eat a conventional or organic diet with identical quantities of the same foods for 22 days. Then their diets were switched and the process repeated. The organic diet contained almost twice the quercetin

* These schools are founded on the ideas and teachings of Rudolf Steiner, who advocated eating food grown using biodynamic practices, many of which support and protect beneficial communities of soil life.

and kaempferol, and those consuming an organic diet had significantly higher levels of both of these flavonoids. Those on the organic diet also displayed increased antioxidant activity. The study, however, did not control for different crop varieties. So the higher level of flavonoids from an organic diet could reflect either differences between organic and conventional cultivars or the effects of growing conditions. Nonetheless, this study demonstrated that higher dietary intake of flavonoids leads to more antioxidants that deliver health benefits.

Further evidence of a link between soil health, farming practices, and human health came in an interesting 2016 comparison in the *New England Journal of Medicine.* Researchers examined asthma rates and immune cell characteristics among Amish and Hutterite schoolchildren of similar ancestry but radically different lifestyles. Amish children lived in communities that used traditional, animal-powered farming practices, whereas Hutterite children lived in communities that employed industrialized, chemical-intensive practices. The researchers found profound differences between the two groups in the type and abundance of immune cells, with the Amish children generally having fewer of the type with proinflammatory action. This difference helped explain why asthma and allergic sensitization were four and six times lower among Amish children, despite seven times the allergen levels in their households.

Why such stark differences? The researchers attributed lower levels of asthma and allergies among the Amish to their early-in-life contact with a diverse community of soil organisms, which led to another effect. Greater exposure to beneficial and neutral bacteria, fungi, and other microbes in infancy and childhood translates into more tolerant immune responses later in life.

Children living on farms also composed one of the groups that researchers analyzed in the multicountry study of allergy incidence among children attending Steiner schools and public schools discussed earlier. Of the three groups, farm kids came out on top with even less incidence of allergies than the Steiner schoolkids.

Finally, a 2018 study of almost 70,000 French adults ranked participants based on their consumption of organic foods and tracked their medical records for an average of almost five years. Greater organic food consumption was associated with higher intake of fruits and vegetables and lower intake of processed meat, poultry, and dairy products. Higher scores also were associated with a lower overall risk of cancer, with those consuming the most organic food having roughly a quarter to a third lower risk than those consuming the least. However, because those who ate more organic food also had higher income, educational level, and physical activity, the differential health outcomes documented by the study could potentially reflect other lifestyle factors.

For now, we know more about the downside of soil influencing our health—about pathogens and contaminants—than about how much human health might improve or chronic diseases decline through improving soil health. This points to the need for a broader perspective on soils and more comprehensive assessments of probable connections between soil health and human health. At the same time, it is also important to remember that other factors, from genome to gender, also affect human health. Still, it remains clear that rebuilding soil health on farmland holds untapped potential that could prove transformative for the way we think about and practice both agriculture and medicine.

Running on Empty

The importance of sufficient levels of micronutrients, beneficial fats, and phytochemicals in the human diet reflects their fundamental importance to our health. What do micronutrients do for us, and what happens to our health when we don't get enough of the right ones?

Micronutrients play leading roles in many aspects of human health and biology. Adequate iron during fetal development and early in life is essential for developing normal cognitive functions. And copper is critical for a properly working immune system, with deficiencies potentially playing a role in Alzheimer's disease. Vitamins B_6 and B_{12} are involved in making neurotransmitters essential to normal brain function, and the latter can delay dementia—provided it's administered before onset of symptoms. Antioxidant and anti-inflammatory phytochemicals help prevent or slow the accumulation of cellular damage that can lead to more serious health problems. Dietary deficiencies in vitamins, minerals, and antioxidants are considered a contributing factor for a host of aging-related diseases.

Soil characteristics, including mineral composition, pH, and organic matter levels, are among the factors associated with human nutrient deficiencies. For many minerals, geology is the largest influence, as it dictates their overall availability in soils.

Iodine became the first micronutrient recognized as essential for human health in 1850 when French chemist Gaspard Chatin found that people living near the sea did not suffer from goiter as much as those who lived in the Alps. He attributed the difference to inland soils being low in iodine, which also meant levels were low in the crops and animals from these regions. For decades his colleagues thought him crazy, until one of them showed that iodine was essential for proper functioning of the thyroid gland. We now know that while we don't need much iodine, we need that little bit quite a lot. Iodine deficiency during pregnancy can even lead to the severe developmental impairment once known as cretinism, but now diagnosed as congenital hypothyroidism. Nutritional synergies come into play with iodine as well. A diet of foods providing insufficient vitamin A and iron can exacerbate the effects of iodine deficiency.

Once thought of only as a likely carcinogen, selenium is now understood to be a critical micronutrient with antioxidant functions important for a healthy thyroid gland and immune system. Low levels of selenium also have been implicated in heart and degenerative diseases. Not surprisingly, selenium deficiencies develop in people who consume crops grown in soils with low levels of it. In yet another illustration of the value of a diverse diet, Brazil nuts happen to be rich in selenium. Several a day can provide all you need.

As a key cofactor or catalyst in thousands of physiologically important reactions, manganese also plays a major role in human health. What happens when you run low? You don't run as well. Manganese is central to minimizing the harmful effects of mitochondria chugging away burning oxygen inside cells. Without adequate manganese, they can fry in their own cellular exhaust. These manganese deficiency–related effects are especially concerning given that glyphosate appears to lock up this mineral element in particular.

Zinc is another important mineral micronutrient for which dietary deficiency leads to major adverse health effects—and, as we've covered, inadequate dietary intake is common. Zinc supports a wide variety of metabolic functions, facilitates protein synthesis, and influences gene expression. Like manganese, zinc is a potent antioxidant. It plays a role in cell growth and stabilizes cell membranes. Long-term deprivation increases oxidative damage and reduces enzyme activity important for maintaining health. The immune and digestive systems, as well as prostate and salivary glands, use zinc for signaling. And as if all these roles weren't enough, it is involved in the synthesis of DNA and RNA and the perception of taste and smell. Zinc also affects the availability of other dietary minerals. It enhances iron absorption, which is vital to avoid stunted growth, fetal abnormalities, delayed adolescence, impaired cognitive capacity, and learning and behavioral issues. All in all, getting enough zinc is paramount for health.

Declining micronutrient levels in crops are of serious concern because more than three billion people experience deficiencies. Recall too that zinc deficiency was virtually unknown before the introduction of high-yielding Green Revolution crops and the fertilizer-intensive practices for which they were bred. Magnesium deficiency is also now common in the United States, with less than half of Americans consuming an adequate amount.

Vitamin deficiencies abound as well, with nine out of ten Americans not getting enough vitamin E through foods and almost a third getting inadequate vitamin C. While supplements can address some deficiencies, so could increasing the vitamin and mineral content of crops through breeding and improved soil health. But, of course, breeding to increase these nutrients won't help much if, as is the case with grains, we then turn around and remove nutrients through processing.

What does a body do when it does not receive adequate supplies of minerals—or for that matter other compounds in food that benefit health and longevity? The triage theory of nutrition holds that even though a person may not meet the standard of deficiency for a given nutrient, shortages still undermine health. Certain vitamins, minerals, fats (omega-3s), and other naturally occurring compounds in food are subject to triage, meaning that when faced with limited amounts, our bodies prioritize them for short-term survival rather then longer-term aspects of health.* It makes sense that natural selection should favor physiological processes related to immediate survival—especially getting to and through reproduction. We all can weather occasional shortages of triage nutrients, but sustained over a lifetime this situation manifests as chronic diseases rather than obvious and visible classic deficiency diseases. In other words, modest deficiencies in micronutrient and other health-imbuing compounds don't affect day-to-day operations as much as they shortchange essential cellular cleanup and maintenance over a lifetime.

While genetics plays a role in human longevity, aging is not programmed from birth. Adequate long-term levels of nutrients subject to triage are essential for minimizing or reversing oxidative damage to DNA, mitochondria, and cellular membranes common to many age-

* So far, nutrients proposed to be subject to triage include minerals (magnesium, calcium, selenium, zinc, iron, and copper), vitamins (B_2, B_5, B_6, B_7, B_{12}, and D), and omega-3s.

related degenerative diseases. So long as maintenance and repair of cells and tissues can keep up with damage, one stands a greater chance of staying healthy while aging.

Many studies have addressed how cultivation practices affect the composition of particular nutritional components of crops, but only a few have tried to address the net impact on public health. After all, there are lots of confounding factors. Crop varieties matter, as do the mineral composition and health of the soil where they grow. Moreover, few people eat either a purely organic or conventional diet.

Still, we know the diets that best support health are those with an abundance and diversity of whole foods. Plants of course are sources of phytochemicals, while ruminant meat and milk deliver fats like CLA you'll never find in a plant food. And whether from plants or animals different foods have a strong suit in terms of particular micronutrients, like vitamin C in citrus and vitamin B_{12} in meat. The winning combination, as you realize by now, is plant foods grown in healthy soil and animals raised on a diet of living plants.

A 2020 study from the Harvard School of Public Health that tracked the diets of more than 200,000 men and women for over 30 years found that proinflammatory diets were associated with an almost 40 percent greater risk of cardiovascular disease. So how do we counter this risk? What makes for an anti-inflammatory diet? Whole foods rich in phytochemicals, fiber, and omega-3s—especially those grown in healthy, fertile soil.

Medical research increasingly supports the power of healthier diets. One pioneering study in Philadelphia found that home delivery of three healthy meals a day for six months reduced health care costs by almost a third for a cohort of chronically ill Medicaid patients. Similarly, researchers at UC San Francisco found that when people with HIV or diabetes who had little access to healthy food were provided three healthy meals a day for six months, they reported decreased depression and greater adherence to retroviral therapy and diabetes self-management. Providing meals for each patient cost half as much as a single day in the hospital, and over the course of the study, progressively fewer patients required hospitalization. All in all, the study demonstrated the positive role diet can play in both the quality of life for people with chronic disease and the potential for substantial savings on medical care.

Going a step further, a local Pennsylvania hospital partnered with the Rodale Institute and started a five-acre organic farm on the hospital grounds to supply food for the patients. Growing recognition of the connections between healthy soil, healthy crops, and healthy people is fulfilling Eve Balfour's vision of getting soil scientists and doctors working together—only decades later than she'd hoped.

Fungal Feats

Our understanding of the roles that fungi in particular play in agriculture has come a long way since Balfour's time. She, Howard, and others grasped that this rather odd form of life—neither plant, nor animal—was, on the whole, more beneficial than harmful to crops. Today we know fungal symbionts that establish relationships with living plants present potential sources of new antibiotics, antioxidants, anticancer agents, and natural insecticides. For example, like their arboreal host, fungi associated with yew trees can produce taxol, a compound used to treat cancer that can deter and in some cases stop tumor growth. A 2021 review of observational studies published since the mid-1960s even found that those who consumed the most mushrooms had a one-third lower overall risk of cancer than those who ate the least.

One of the most beneficial compounds in mushrooms was only discovered because of a problem fungus, the ergot producer *Claviceps purpurea*, which spends part of its life cycle in the soil before infecting grasses like wheat and rye. This fungus not only lowers yields and reduces grain quality; it makes people hallucinate when they eat grains from infected plants thanks to a compound related to lysergic acid, the precursor for LSD. Various events throughout history have been ascribed to ergot-tainted grain. Most well known is Saint Anthony's fire, a common illness in medieval times in which people suffered from hallucinations and burning sensations in their limbs. Ergot-tainted grain is also linked to an outbreak of violent hallucinations in southern France after the Second World War.

But there is another connection involving agricultural practices and human health involving this same ergot-producing fungus. In 1909 French pharmacist Charles Tanret reported isolating a mysterious amino acid, ergothioneine (or ergo for short), from the fungus. Half a century later, researchers found the amino acid in almost all plant and animal cells. Crops took it up from the soil, yet the source remained unknown.

Since Tanret's original discovery, other soil-dwelling fungi and certain bacteria have been found to produce ergo. We know, for example, that levels of ergo in some medicinal plants depend on symbiotic soil fungi. And recent research points to the type and abundance of microbes in the soil as key factors in setting ergo levels in food. As our bodies cannot make ergo, we depend on dietary sources to get it. While animal foods are also a good source, garlic and mushrooms offer the most concentrated sources of ergo, although levels in different kinds of mushrooms vary tremendously.

Ergo is a good example of how the conventional framing of a field of study—in this case nutrition—can impede understanding of new advances in science and integrating them into practice. As ergo is not critical for growth or survival, it does not count as a nutrient under standard definitions, nor does it have recommended intake levels like those for vitamins and minerals. Yet there is strong evidence that ergo has substantial health benefits. Human cells have a highly selective membrane transporter that detects and moves the compound out of the bloodstream and into cells, implying a long-running evolutionary benefit.

So just what is ergo doing inside the cells of our bodies? It is one of a small handful of compounds central to healthy aging proposed as "longevity vitamins" (though not all are vitamins). The idea is that when cells are deprived of longevity vitamins, the proteins and enzymes that prevent and fix cellular injuries become less active, allowing damage to accumulate. Ergo is considered a candidate as it prevents ongoing stress and cellular damage, including in the DNA of mitochondria. In other words, ergo and other longevity vitamins control and modulate protective and repair mechanisms inside of our cells that help keep them running smoothly.

Key evidence supporting the importance of ergo comes from experiments on human cell lines in which removing the membrane transporter that moves ergo into cells led to low ergo levels and subsequent oxidative damage to proteins, lipids, and DNA. The outcome? Most of the cells were so damaged that they died. Such oxidative damage is characteristic of chronic age-related degenerative diseases.

Additional evidence that implicates low ergo levels in disease onset or progression comes from a Japanese study that reported much more ergo in the blood of healthy volunteers than in patients with Crohn's or Parkinson's disease. Other studies report that ergo exhibits protective effects against cardiovascular and neurodegenerative diseases. A Swedish study that followed several thousand middle-aged participants for over 20 years found that out of more than 100 measured metabolites, higher ergo levels were most strongly associated with lower risk of coronary disease and mortality. On the other side of the world, a study of elderly people in Singapore found that while blood levels of ergo decline as people age, cognitively impaired individuals over age 60 had the lowest levels. Greater estimated annual consumption of ergo in the United States and several European countries (Finland, France, Ireland, and Italy) is also correlated with increased longevity and lower rates of Alzheimer's and Parkinson's.

While such correlations do not establish causality, the foregoing does point to a connection that charts a direct line from soil health to human health. It stands to reason that agricultural practices that reduce fungal and bacterial abundance and diversity in soils likewise influence ergo levels in food. The connection is not so far-fetched. In 2021, Penn State researchers published a study that investigated how different tillage methods impacted ergo levels in three crops (corn, soybeans, and oats). Experimental plots for each crop received one of three treatments: intensive tillage (moldboard plow), minimal tillage (chisel disc), or no plowing (no-till). For all three crops, those grown in the no-till plots had the highest ergo levels.

Joining ergo on the roster of proposed longevity vitamins are pyrroloquinoline quinone (thankfully shortened to PQQ) and queuine. Rhizosphere-dwelling bacteria make these two compounds, and crops take them up from the soil. Plants use PQQ as a growth hormone, and queuine is part of the elaborate dance nitrogen-fixing bacteria engage in to enter a plant's root cells. Both compounds also influence other aspects of plant physiology, and it is reasonable to think that tillage and fertilization practices influence their levels in crops and thus how much we and animals ingest in our respective diets.

Queuine and PQQ are further examples of the need to better integrate noncaloric dietary elements into how we define nutrients and think about linkages between human health and diet. Like ergo, they differ from the nutrients that function as energy sources and support growth, yet substantial evidence points to their positive influence on health as people age. In our bodies PQQ is especially important for mitochondria to function normally, and it appears to act as a powerful antioxidant that helps prevent neurodegenerative conditions, cardiovascular disease, and type-2 diabetes. Queuine influences numerous aspects of cellular function ranging from anticancer effects to increasing the efficacy of certain antioxidants. Research using a mouse model of multiple sclerosis even found that queuine supported cellular activity that led to full remission.

A small handful of carotenoids are also candidates for longevity vitamins.* All function as antioxidants that help protect plants from the ravages of sunlight. In people low levels of these carotenoids are implicated in maladies ranging from macular degeneration–induced blindness to cognitive and immune disorders.

Mycorrhizal fungi, as you now realize, are masters at increasing nutrient uptake and helping to protect plants from pests and pathogens. They also increase levels of phytochemicals beneficial to human health, as shown in studies on tomatoes, chili peppers, strawberries, lettuce, basil, artichokes, onions, and sweet potatoes. But for each crop the effect depends on finding the right fungal partner. Plenty of work remains to be done to identify and foster crop-bacteria-fungus combinations that help mobilize and deliver minerals, boost phytochemical density, and reduce fertilizer, pesticide, and energy use. Increasingly, such opportunities line up with what we now know and understand about the world of nature.

Restoring fungal allies to cropland would also help close the organic yield gap. But to do so we'd have to start breeding crops for association with mycorrhizal fungi—and stop disrupting them with tillage. After all, long, long ago fungi helped plants conquer the continents. Now they can help secure agriculture's future—and our health.

Farming practices contribute to whether soil fungal communities harm or benefit us in other ways too. Less tillage means fewer airborne particulates and less chance for exposure to dust bearing the soil-dwelling fungus that causes Valley fever in California's Central Valley and other agricultural regions in the West. Simply not tilling the soil would rein in this serious public health hazard. Here again, practices that build soil health promote human health.

* These carotenoids include several found in land plants (lutein, zeaxanthin, lycopene, alpha- and beta-carotene, and beta-cryptoxanthin), as well as a marine carotenoid (astaxanthin).

Even introducing children to gardening in schools can help improve diets. A study of middle school students found that those who engage in gardening at school increased their preference for vegetables in general, consumed a greater variety of them, and were more likely to try unfamiliar ones. Likewise, a study involving three broad ethnic groups (Black, White, and Hispanic) in a diverse northeastern U.S. city found across the board that eating fresh-picked fruits and vegetables while growing up or gardening as an adult increased vegetable consumption. Here is another way that soil health influences human health: physical and personal connections to soil and growing food can improve dietary choices. It also grounds kids in understanding that food comes from the soil.

Dietary Diversity

Despite a flood of dietary advice and associated research, there is a remarkably simple way to avoid micronutrient deficiencies. Eat like a human. There is power in the omnivorous diet of whole plant and animal foods we evolved to eat. But we keep chiseling away at the diversity omnivory allows. Almost 200,000 flowering plants produce edible parts. And yet we humans eat fewer than 300 of them. Just a dozen and a half provide almost 90 percent of our food. We've whittled down the breeds of every farm animal too. Over the past century diets around the world simplified under modern industrialized agriculture, with three grains—wheat, corn, and rice—now accounting for more of what people eat than the two dozen next-most-consumed crops. In some developing regions a single grain (rice or wheat) now provides more than 80 percent of daily caloric intake.

The 2019 EAT-Lancet Commission report on how to sustainably feed humanity cast the current global food system as detrimental to both human and environmental health. This commission of 37 experts from 16 countries advocated reorienting the food system around producing healthy food rather than high yields. Concluding that unhealthy diets pose a greater risk to health and well-being than drug, alcohol, and tobacco use combined, they also pointed to conventional food production as a primary driver of climate change, biodiversity loss, and pollution. The commission recommended a global diet rich in a diversity of plant foods, including doubling consumption of fruits, vegetables, nuts, and legumes. They also advocated eating modest amounts of fish or poultry and far less red meat and dairy. That sounds to us like an omnivorous diet of whole, fresh, minimally processed foods—the diet that supported human health throughout our evolution.

The recommendations related to poultry, red meat, and dairy consumption did not, however, distinguish between grain-fed animals raised in confinement and those raised in settings and on diets that lead to dramatically different—and better—phytochemical and fat profiles as well as significant potential for environmental improvements. Nor did it address how crops are grown.

And yet the commission concluded that a revolution in agricultural practices was needed to close yield gaps, greatly reduce nitrogen fertilizer use, and shift agriculture from a carbon source to a carbon sink by increasing soil organic matter. How could all this be done? Adopting agricultural practices that rebuild healthy, fertile soils.*

While Hippocrates's famous advice to use food as medicine still makes good sense in terms of preventive medicine, it increasingly seems that to harvest the fullest benefits we need to pay greater attention to the essential dimension of how that food is grown—to how we farm.

* Across the board, there are opportunities in both conventional and organic farming systems to improve and maintain soil health.

HARVESTING HEALTH

If we destroy our soil, mankind will vanish
from the earth as surely as the dinosaurs.

—EVE BALFOUR

Laying out the evidence connecting soil health to crop health, livestock health, and human health, the pieces fit together to reveal a fundamental truth. Healthy, life-filled soil makes for healthy, nutritious crops, forage, and livestock that, in turn, support human health. Seen in this light, our health or lack of it reflects how we treat the land.

Are modern farming and dietary changes to blame for the modern epidemic of chronic diseases? Not entirely. Huge shifts in chemical exposure, physical activity, and lifestyle affected human populations over the same time frame. But changes in how we grow what we eat inadvertently undermined our health.

It turns out that the hallmarks of agriculture today—breeding for yield, liberal use of synthetic fertilizers, an arsenal of biocides, and creating diets and environments for farm animals ill-matched to their biology—delivered outcomes we didn't expect. Chasing yield above all normalized farming practices that killed off beneficial soil life and disrupted nutrient cycling in ways that lower the amount and change the mix of phytochemicals, micronutrients, and beneficial fats in our food, none of which helped us. Confining cows and other farm animals to indoor settings while shifting their diet from living plants to TMRs reduced beneficial fats in meat, eggs, and dairy before most of us ever heard of such a thing.

How much did all this matter? The reality today is that seven out of ten Americans live with chronic conditions uncommon until a few generations ago. Something's not working.

The importance of sufficient dietary calories to grow and develop normally from birth into young adulthood is unquestionable. But there is more to health than reaching the milestone of physical maturity. For in the decades that follow, if part of us falters long enough to seed ill health, the whole suffers. Antioxidants and anti-inflammatory compounds in the foods we eat are crucial in this regard. Like a Swiss army knife for our bodies, they provide a basic array of tools to fix and repair small problems before they turn into more serious or chronic ailments. A well-provisioned diet helps keep our bodies humming along as we age. In this regard, certain fats are surprising must-haves for maintaining the immune system's turn-on-a-dime capacity to manage inflammation, dialing it up and back down as needed.

It is not unusual for theory and practice in major fields of human endeavor to carry baggage from the past that can impede accepting new insights and knowledge. After germ theory emerged as a guiding paradigm in microbiology, it blinded medicine and agriculture to the foundational role of symbiotic relationships with microbial life and we set out to sterilize our bodies and farms.

A simplistic, single-factor paradigm also carried over to the founding of nutrition as a discipline. Once we learned that a single nutrient could cure a case of scurvy or rickets, the hunt was on to identify more one-to-one causal linkages between diet and disease. But this model doesn't work for most chronic diseases. It neglects the importance of dietary synergies—like omega-3s packaged with phytochemicals in grass-fed meat and dairy and the thousands of phytochemicals in whole plant foods. And to function best, numerous aspects of human physiology require noncaloric compounds relegated to "non-nutrient" status or saddled with wonkier terms like "bioactive." Ergothionene, PQQ, and queuine are the proverbial tip of the iceberg in this regard. Agriculture and nutrition alike need to cast off outdated ideas inherited from their historical roots.

We notice health most when it is lacking—in both people and the soil. This is where farming practices come in. Those that build soil health can heal the land rather than rob it of its lifeblood—organic matter and microbially driven nutrient cycling. Regenerative farming offers a recipe for harvesting quantity *and* quality. It also can broaden how we think about and define nutrients.

A diverse array of whole foods that engage our body wisdom is the time-tested way to meet our body's needs. With such a diet our cells, like the roots of a plant in healthy soil, take up things that fuel growth early in life, as well as phytochemicals, micronutrients, and recently recognized longevity vitamins vital for good health throughout life. In plain terms, every body's cells must pull in energy, take out the cellular trash, and do the upkeep needed to fend off illness and disease. This is why farming practices matter to human health: they can imbue the foods in the human diet with the intelligence on which body wisdom feeds. In short, agricultural practices can and should play a major role in providing foods that better support disease prevention throughout life.

In 2018, health care in the United States cost $3.6 trillion, or about $11,000 per person. This is about twice as much per person as in other developed countries. A 2021 report from the Rockefeller Foundation found that the diet-related human health costs in America roughly equaled the cost of food, making today's historically cheap food not such a good deal. Supporting farmers who grow crops and raise livestock in ways that enhance nutrient density would pay dividends in helping prevent chronic diseases and lower health care costs down the road.

New Direction

When it comes to health there really is no need to strive for some sort of super status. For plants and animals alike, including us, health is the normal state and sickness the abnormal one. With health comes resilience, the ability to bounce back rather than collapse when facing disease or illness.

Boring as it may sound, cells, tissues, and organs that function normally set us up for good health. Yet normal is different for everyone due to our unique biochemistry and genes—our own and those of our microbial tenants. These differences make defining a standard diet problematic, especially for countries like the United States where modern populations reflect a history of colonization, slavery, and waves of immigration. Should a person of African ancestry with high blood pressure eat the same diet as a person of Irish descent with a family history of breast cancer? Consider as well Indigenous populations long ignored in dietary and medical research. The diverse backdrop on which the human diet acts is part of why there is no single, "best" diet for us all. Variability is the one true across-the-board constant.

Multiple parallel changes in recent decades complicate establishing causality for the modern epidemic of chronic diseases, yet the roots of many ailments trace back to what we eat and how we farm. Viewed broadly, there is no single food or chemical we can blame. The problem is bigger—conventional agriculture that delivers a harvest of degraded soils and low-nutrient-density food laced with toxins. Fortunately, we have an alternative.

Well-implemented regenerative farming practices can bring our soil back to life, reduce levels of things we don't want in our food, and boost those we do. Despite dietary guidelines not considering how food is grown, a chain of causal mechanisms and evidence nonetheless connects soil health to human health. Beneficial fungi that fetch nutrients and soil bacteria that hand out nitrogen, rat out pathogens, and prompt plants to make phytochemicals are all hallmarks of healthy soils. Their activities imbue crops and livestock with compounds that, once in us, help keep inflammation under control and combat oxidation. Rebuilding soil health is how to pull off growing an abundance and diversity of nutrient-dense foods. Agriculture is long overdue for a reset to focus on cultivating the life we need rather than killing it.

Interest in rethinking how food is grown isn't just a big-city thing. Degraded soil and declining farm profitability motivate the farmer-led soil health movement taking root across the country and around the world. And the regenerative practices used to restore soil and farms also benefit local communities and society at large. Healthier food, less water and air pollution, and more carbon returned to the land to feed soil life add value to the harvest.

Unfortunately, the average consumer doesn't usually have the point-of-purchase information they would need to support farming practices that build soil health. The familiar organic label can fall short in this regard. But that's starting to change. Much like long-established permaculture and biodynamic practices, new certification programs focused on soil building go beyond the USDA organic label.*

We need farmers of all stripes, conventional and organic alike, to adopt practices that build soil health and thereby cultivate human health. In aiming to transform conventional agriculture, it's useful to think of regenerative farming as an umbrella covering practices that build soil health, including organic-ish farmers who minimize chemi-

* These include the recently introduced "regenerative organic" and "real organic" labels.

cal use without necessarily forgoing it altogether. While some controversy surrounds how to define regenerative farming, we see its defining trait as incorporating practices that build and maintain fertility as a *normal consequence* of farming.

Indeed, momentum is growing for a revolutionary agricultural realignment as practices centered on soil health catch on in both organic and conventional camps. Systems that combine minimal disturbance (no- or low-till), continuous growing biomass (cover and companion crops), and diversified harvests and rotations offer flexible, adaptable ways to use less diesel, fertilizer, and pesticide. And farmers who use less of these things keep more money in their pocket. They tend to like that.

Regenerative grazing practices can deliver an enhanced phytochemical content and healthier fat profile in meat, eggs, and dairy. Consumers should like that. Unfortunately, studies that assess the effects of meat and dairy on human and environmental health generally do not consider, let alone account for, differences in animal diets. Reducing the overall consumption of meat from feedlot-raised animals while increasing the proportion of livestock grazing diverse, phytochemical-rich pastures would improve the health of livestock, people, and the land. Needless to say, there's a lot of room for positive change.

The basic choice we face around meat isn't so much all or none but how much we eat and how to raise it. In this regard, farms like Jonathan Lundgren's and Gabe Brown's provide models that meet the fundamental freedoms of farm animals while restoring health to the land. As research on the net climate impact of grass-fed versus feedlot beef continues to unfold, the factors to consider include the style of grazing, the type and diversity of forages, the use (or not) of synthetic fertilizers on pastures, and the energy inputs for all the supply chains needed to make feedlots possible. Reorienting American beef production around consuming less, yet more-nutritious meat would reduce agriculture's environmental footprint. It could benefit our health too.

Assessing climate impacts from animal agriculture also depends on whether pastures occupy deforested or restored prairie. For example, a 2020 assessment of a rotational grazing system running chickens, cattle, sheep, and pigs on restored pasture in Georgia found that including soil carbon sequestration in a full life-cycle assessment reduced greenhouse gas emissions by 80 percent and reduced the overall carbon footprint by about a third compared to conventional production. While this diversified production system required somewhat more land, it restored degraded grazing lands and converted land once used to grow feed corn into diverse pastures that better sequester carbon and protect biodiversity.

In addition, an important, often-neglected aspect of animal agriculture is that methane emissions from ruminants can drop sharply when they consume a diet of common forage plants rich in tannins and saponins. How so? These phytochemicals reduce populations of rumen microbiota that produce methane. In other words, grazing cattle on a diverse diet of grasses, forbs, and shrubs can reduce the carbon footprint of meat and dairy products.

But so far the opportunity to park carbon in the soil is not the primary reason behind why practices that improve soil health are catching on for raising livestock and growing crops. Increasingly, they make economic sense for farmers facing financial crises rooted in modern agriculture's reliance on costly inputs. Ditching the plow reduces labor and fuel use, saving time and money. Farmers can also cut their expenses for chemical fertilizers and pesticides, in some cases by more than half. Yet yields typically remain comparable or even increase as soil health improves. So for more and more farmers it boils down to simple math, along with some faith and a little patience as their soil heals and their bank account grows.

The only ones who really lose when the soil comes back to life are those making and selling agrochemical inputs to farmers. Who gains? Everyone else.

Regenerative practices are gradually catching on. In 1990 no-till farmers worked about 6 percent of U.S. farmland. Today, around a quarter of American cropland is under continuous no-till, but just a fraction of that also uses the other two key practices—cover crops and diverse rotations. Global acreage under conservation agriculture using all three practices grew from fewer than 8 million acres (the size of Maryland) in the early 1970s to almost 450 million acres (the combined size of Alaska and California) by 2016—roughly a tenth of the world's cropland. As of 2018 the annual global acreage converted to conservation agriculture amounted to some 26 million acres, about the size of Ohio. These on-the-ground changes attest to the feasibility and practicality of shifting mainstream agriculture to methods that improve rather than degrade soil and land.

Even with adoption taking off, it will take sustained effort to make cropping and grazing practices that build and maintain soil health the conventional methods of the future. And though the guiding principles appear to be universal, how specific practices get implemented varies for different landscapes, climates, and crops. Still, a common foundation would be to reestablish partnerships between crops and mycorrhizal fungi and to breed for nutrition, resilience, and performance in regenerative systems. It would help to build markets for a greater diversity of food crops and investigate how region-specific rotations and diverse mixtures of crops affect overall system performance. We need to stop routine plowing, start retaining crop residues, and expand the use of microbial inoculants, rotational grazing, and composting practices. Also much needed are strategies like integrated pest management to reduce pesticide use. All of this will take creativity, ingenuity, and support at multiple levels.

Several things already stand out. Rebuilding soil health is not only possible; it can happen relatively quickly using minimal to no chemical inputs. And doing so can help address the long-running and linked problems of declining soil fertility and human health. Moreover, regenerative practices that move carbon from the atmosphere into the soil turn this global threat into a societal asset.

No one can reasonably argue that the high yields of modern agriculture didn't increase humanity's access to adequate calories. But as we've emphasized, the food we eat does not just fuel the growth that transformed each of us from helpless babe to strapping adult. We need our diet to support and equip our bodies with all they need over life's long haul.

COVID-19 laid bare the vulnerabilities of those with chronic diseases in the midst of a pandemic. Another sort of vulnerability came to light when America's meat-packing plants had to shut down, exposing the fragility of industrialized animal husbandry. But the pandemic also showed how resilient—and adaptable—diversified, smaller-scale supply chains can prove. When the closure of restaurants eliminated wholesale accounts for small farms, many shifted to direct-to-consumer sales and e-commerce. That's not to say it was easy or that everyone succeeded. Still, the lesson matters. Distributed systems of smaller, interconnected pieces can react quickly, making them more resilient than centralized industrial systems too rigid to rapidly respond to crises.

Diversifying crops can also enhance resilience at the scale of individual farms. Growing just one or two crops leaves a farmer vulnerable to tariffs and other marketplace disruptions. Integrating a greater diversity of crops into conventional rotations could involve including more legumes and fiber crops (like flax and hemp) along with more conventional grains and produce. If farmers routinely rotated four or more crops in their fields using regenerative methods the land would be more resilient as well. Organic matter–rich soils retain moisture better, helping sustain crops through droughts.

Greater crop diversity could map right onto the food, fiber, and fodder we need farms to produce. Over the past half century American farms increasingly specialized in single crops or two-crop rotations like corn and soybeans. While growing more diverse rotations can help heal the land, farmers of course need to sell what they grow.

And this requires agricultural, distribution, and consumer infrastructure to enable them to raise and market a greater diversity of crops and livestock. If all they can sell in their area are corn and soybeans, that's what they'll grow. And for farmers to grow and sell a greater variety of crops, we the people need to eat a more diverse diet.

Our current food system was shaped over generations by policies and socioeconomic pressures. While economic drivers are starting to work in favor of regenerative practices, our policies and subsidies are not—at least not yet. They continue to support the agriculture of yesterday when we need to transition to the farming of tomorrow. Encouraging farmers to adopt practices that build soil health could employ tax incentives or carbon credits. Doesn't it make more sense to subsidize growing crops in ways that help us grow even more in the future rather than the other way around as we do at present?

But there is another, underrecognized reason to reform our agricultural system—our health. It's time to consider farming practices and the agricultural policies that drive them as critical and overlooked parts of public health and health care. Healing the land and restoring our own innate resilience and disease-fighting capabilities to reclaim our health ultimately rest on the same thing—healthy soil. And to get there we need a unified theory of agriculture and nutrition built on a common foundation of soil health to deliver prosperity to farmers and well-being to eaters. Agricultural policy is health policy.

Future Farming

What could support for a soil-health-centric style of agriculture look like? Consider how we might design policies and invest in infrastructure tailored to three scales of farms. Large grazing or cropping operations could be supported through tax and carbon credits, reforming subsidies, and building markets for a wider variety of crops. Farms and ranches larger than a thousand acres are well suited for reintegrating livestock and commodity crop production. More traditional family farms that are tens to hundreds of acres in size also could be incentivized to reengage in mixed cropping and animal husbandry. In addition, we need to foster small-scale, no-till vegetable farms in and around cities. From less than an acre to 20 acres in size, they could repurpose urban organic wastes to build soil fertility and provide fresh food close to urban markets.

Perhaps we need a new Homestead Act offering young farmers sweat equity loans to rebuild soil fertility on degraded farmland. A visionary investment in America could inspire a new wave of farmers working smaller, more profitable regenerative farms, leading to the revival of rural towns and economies. Farms that support rural communities in this way would help address climate change and reduce off-site pollution from agricultural runoff, things that would benefit us all.

Imagine too, farmers getting paid for the nutrient density and levels of health-promoting and protective dietary components in crops, meat, and milk—or for simply adopting practices known to build and maintain soil health. Over the past century, governments incentivized and subsidized high yields because it translated into a lot of cheap calories. Yet health is not a product, something you can buy in a box or a bottle. To harvest the full benefits rooted in soil health, farming needs a dual paradigm. Put simply, we should be subsidizing farming practices that deliver nutrient-dense food—both quantity *and* quality. Such an imperative recognizes that the interests of individuals, society at large, and the environment all align through connecting soil health to human health. Seeing soil as a trust held for future generations would help us create economic institutions and arrangements that reward farmers for improving their soil rather than subsidizing conventional practices that degrade soil health.

Farming can either mine soil and its fertility or build and sustain soil health. And what we do to the soil this century will affect humanity for centuries to come. If we continue to focus narrowly on growing high yields at the expense of soil health and nutrient density, we risk solving hunger without securing health. Continuing down this path would repeat mistakes of prior civilizations, squandering humanity's greatest natural assets—healthy fertile soil and a healthy populace. The estimated economic value of soils comes to at least $20 trillion, about the size of the U.S. economy, making soil the most valuable natural resource in the world. Beyond irreplaceable, when we take care of it, soil takes care of us.

Because the processes driving soil degradation arise from social, economic, and political forces, rebuilding healthy, fertile soils to nourish the future needs to be rooted in all three as well. When desperate people pass their suffering on to the land, degraded soil hands impoverishment right back. There is a compelling need for a global shift in how we see the soil, to accept the health of the land as central to our own.

We don't face a dire choice between feeding people today and improving soil for tomorrow. As we've discussed in earlier chapters, higher levels of soil organic matter positively influence crop yields. On this point, a 2019 global analysis from the Yale School of Forestry that accounted for factors like nitrogen fertilizer use, soil type, and climate found that increasing organic matter levels raised crop yields for soils with less than about 4 percent organic matter. With much of the world's agricultural soil now below half that amount, there's a lot of room for improvement—increasing soil organic matter from 1 to 4 percent could reduce fertilizer use by almost three-quarters. Rebuilding soil organic matter would not only take carbon from the atmosphere and stash it belowground; it could greatly reduce synthetic nitrogen fertilizer use.

We need to move beyond familiar arguments over conventional versus organic farming to focus on soil health and the key biological processes that help move nutrients and other beneficial compounds into crops. Yet what works in Kansas or California will differ from what works in Ghana or India. Embracing general principles of regenerative agriculture and adapting practices to a farm's specific setting while engaging farmer ingenuity is what, we believe, will largely determine the pace and success of returning health to the land. The bottom line is that building soil health is the right path forward.

As a matter of routine use in both medical and agricultural settings, practices tend to get downplayed for a simple reason. They cannot be as readily marketed and sold as products like the fertilizers, insecticides, and herbicides that reliably deliver hefty profits for dealers. Likewise, the medical world devotes more research to developing new drugs than to health-promoting practices. It is harder to sustain handsome profits when new practices solve problems or prevent them from occurring in the first place.

Yet behavioral changes that drive new practices can support new technologies and products, like no-till planters and mobile fencing for livestock. And farmers developing product lines of cover crop seeds for regenerative farms show how changing practices can lead to new products. We need both approaches—new practices and products—to promote building soil fertility and bringing life back to our fields to benefit crops, farmers, taxpayers, and consumers alike.

Choosing Life

We already use between a third and half of our planet's ice-free land for cropping and grazing. So what happens on farmland will shape biodiversity and the future of nature, including humanity. In the broadest view, a philosophy of cultivating and caring for life is the foundation for a soil ethic. Soil not only feeds us today but will feed our children's children, and this calls for some reverence. Grounding farming in a mindset based on cultivating beneficial life belowground would take care of our tiniest brethren, the herds of soil-dwelling microbes that have long proved essential to the well-being of all plants. And what goes for the smallest creatures applies equally to the largest. Adopting animal welfare as a foundational tenet of animal husbandry, where herds move and eat as they evolved to, is better for our bodies too.

The idea of improving soil health as a way to help improve human health remains virtually unexplored in diet-related medical and nutritional research. Yet how we treat the land will grow the fibers we weave into a future. Will it be moth-eaten and tattered before it makes it off the loom? Or will investing in the health of the land and people help strengthen the fabric and produce a more resilient world? We still have time to choose the regenerative path for our soils, our planet, and ourselves. We just can't afford to wait much longer.

Yes, it really does matter how we grow our food. We need to stop running from the messy, complex reality that what's good for the land is good for us too. We seem to keep relearning this age-old lesson and missing the one Eve Balfour recognized almost a century ago—that how we farm shapes what we become. In the end we all need to face the simple, bite-sized truth—you are what your food ate.

ACKNOWLEDGMENTS

Writing a book can feel like a lonely marathon, but we are fortunate to have a great team backing us up and shepherding us along. First, we thank The Dillon Family Foundation for supporting both the nutrient density analyses and time to write, without which this book would not have happened. Our agent, Elizabeth Wales, once again helped us frame and launch the book. Her insights, as always, better shaped our arguments. We thank the W. W. Norton team, especially our editor, Melanie Tortoroli. Responsive and quick-witted, her enthusiasm and guidance buoyed us along. We are particularly grateful for her patience and suggestions in helping us craft and streamline the manuscript. Carina Wells helped with library research, and many others too numerous to name shared useful references and ideas. And we're grateful to Gustavus Adolphus College for providing a welcoming place to write some of the early chapters.

We thank Garrett Duyck and Noah Williams for alerting us to the potential for testing wheat from their cover crop trial—and for sending us samples. We also thank Alex Gagnon and Tamas Ugrai, respectively, for use of and help in the lab for running high-resolution atomic mass spectrometry to determine the mineral content of our initial wheat analyses. Alexander Michels and the staff at the Linus Pauling Institute at Oregon State University were tremendously helpful in guiding and conducting the analysis of crops from farms we sampled. Ray Archuleta was not only incredibly helpful and insightful in accommodating the soil-sampling desires of a pair of writers but a lot of fun to travel with along the way. Steve Jones shared samples of wheat, and Gabe and Paul Brown shared their test results. Bryan and Anita O'Hara, Paul and Elizabeth Kaiser, Steve Jones, Jonathan Lundgren, and David Johnson and Hui-Chun Su Johnson were delightfully accommodating of our visits. Norman Uphoff shared some of the history on SRI rice and helped guide us to research on the topic.

Finally, we are deeply thankful to all the farmers, ranchers, and researchers we met along the way who gave generously of their time, showed us their farms, and allowed us to sample their soils and crops. Naturally, any inadvertent errors that come with writing a book like this remain ours alone.

SOURCES

We relied heavily on articles in peer-reviewed journals as well as books in researching *What Your Food Ate.* As our source list contains around a thousand references and runs to over fifty pages, we opted for an e-bibliography to save trees and reduce the size, cost, and weight of the book. For readers interested in digging deeper, the source list is available at www.dig2grow.com.

INDEX